China's Energy Security

A secure supply of energy is essential for all nations, to sustain their economy, and indeed their very survival. This subject is especially important in the case of China, as China's booming economy and consequent demand for energy is affecting the whole world, and in turn potentially driving realignments in international relations. Moreover, as this book argues, energy security should be considered more broadly, to include issues of sustainability, environmental protection and the domestic organisation of energy policy and energy supply. This book presents a comprehensive picture of China's energy security. It covers all energy sectors – coal, oil, gas, renewables; international relations with all major sources of energy supply – the Middle East, Central Asia, Africa; and key areas of domestic policy-making and supply.

Giulia C. Romano is the Energy Programme Manager at Asia Centre, Paris, France.

Jean-François Di Meglio is President of Asia Centre, Paris, France.

Routledge Contemporary China Series

For a full list of titles in this series, please visit www.routledge.com.

China's Energy Security

A multidimensional perspective

**Edited by
Giulia C. Romano and
Jean-François Di Meglio**

Routledge
Taylor & Francis Group

LONDON AND NEW YORK

First published 2016 by Routledge

2 Park Square, Milton Park, Abingdon, Oxfordshire OX14 4RN

711 Third Avenue, New York, NY 10017

Routledge is an imprint of the Taylor & Francis Group, an informa business

First issued in paperback 2017

British Library Cataloguing in Publication Data
A catalogue record for this book is available from the British Library

Library of Congress Cataloging in Publication Data
Names: Romano, Giulia C., editor. | Di Meglio, Jean-François, editor.
  Title: China's energy security : a multidimensional perspective / [edited by]
Giulia C. Romano and Jean-François Di Meglio.
  Description: Abingdon, Oxon ; New York, NY : Routledge, 2016. |
  Series: Routledge contemporary China series ; 151 | Includes
bibliographical references and index.
  Identifiers: LCCN 2015047386| ISBN 9781138913141 (hardback) | ISBN
9781315690926 (ebook)
  Subjects: LCSH: Energy security–China. | Energy policy–China. | Energy
development–China. | Energy industries–China.
Classification: LCC HD9502.C62 C49635 2016 | DDC 333.790951–dc23
LC record available at http://lccn.loc.gov/2015047386

ISBN: 978-1-138-91314-1 (hbk)
ISBN: 978-0-8153-5598-4 (pbk)

Typeset in Times New Roman
by Taylor & Francis Books

Contents

List of illustrations

Figures

Tables

Boxes

List of contributors

Giulia C. Romano is researcher at Asia Centre and doctoral researcher at Centre des Recherches Internationales (CERI) Sciences Po, Paris, France.

Jean-François Di Meglio is President of the French think-tank Asia Centre.

Zhi Liu is doctoral researcher at Tsinghua University, China.

Zhe Wang is Postgraduate Student at Renmin University, China.

Na Yin, PhD, is director of the Research Department of Beijing Enterprises Group Company Ltd., Beijing, China.

Xuanxuan Zhang is graduate student of the department of Engineering, Auburn University, AL, USA.

Benjamin Denjean is doctoral researcher at the Beijing Forestry University, College of Nature Conservation, Beijing China and Manager of the Association for Vertical Farming China Chapter.

Cyril Cassisa is visiting scholar at the Institute of Energy, Environment and Economy, Tsinghua University, Beijing, China and President of Shin Development Association, Montpellier, France.

François Bafoil is CNRS Senior Research Fellow at Centre des Recherches Internationales (CERI) Sciences Po, Paris, France.

Michal Meidan is Director at China Matters and Associate Fellow at Chatham House.

Vanessa Boas holds a PhD from the University of Cologne (Chair Jean Monnet) and Charles University, Prague (Centre for Balkan, Eurasian and Central European Studies).

Olga A. Spaiser is doctor in Political Science at the Center for European Studies (CEE) and lecturer at Sciences Po, Paris, Le Havre and Nancy.

Ka-ho Yu is a Postdoctoral Fellow at Harvard Kennedy School and a Research Fellow at Center for International Energy Security Studies, Chinese Academy of Social Science (GS).

Yunheng Zhou is researcher at Environmental and Energy Policy Center, Zhejiang University.

Patrick Schroeder is Key Expert on Sustainable Consumption and Production for the EU's SWITCH-Asia Network Facility, based at the Collaborating Center on Sustainable Consumption and Production (CSCP).

Marie-Hélène Schwoob is a research fellow and project coordinator at the Paris Institute for Sustainable Development and International Relations (Iddri) and a former project manager of Asia Centre's Energy Programme.

Acknowledgement

A small but enthusiastic group of people made the realisation of this book possible and we would like to thank all of them for their precious contribution. Laura and Peter Brown generously helped us in harmonising the language of these 'multi-English' chapters and preciously supported us in the editing process. Without their help this work would have not been possible. We also received a generous and untiring encouragement and assistance from François Bafoil, at CERI-Sciences Po, who also contributed to a chapter. Our cooperation and discussion beyond the preparation of this book on energy issues in Asia and in Europe is one of the best achievements and presents of this joint work and a new step for future collaborations in research on energy topics. *François, un grand merci!* A great thank has to address our team of co-authors, for their precious work and patience and for the extensive exchanges conducted with many of them. Through their great contribution and their different perspectives – coming from various professional and educational backgrounds – we could extend our understanding on energy topics in China beyond our focus on politics and policy implementation. A thank also to Roland Winkler, for which the nexus urbanisation-energy in China has no secrets anymore. And a final special thank to the patient support and help of Jodie Cazau and to the Asia Centre team. This is our common work, our common achievement and we would like to dedicate it to all our team.

Giulia C. Romano and Jean-François Di Meglio,
October 2015

1 Introduction

From 'shaping' to 'framing' China's energy security and the example of the oil policy

Jean-François Di Meglio and Giulia C. Romano

This collective work examines the evolution of China's energy policies over the past decade in the context of a conceptual shift from 'shaping' to 'framing'. It continues the research of the first collective contribution on the topic of energy security in China, *Shaping China's Energy Security* (Meidan 2007). Our prime interest is to understand the behaviour of China as the main actor in world energy markets. This has led us to analyse changes in energy security concepts and actions, from the time of China's 'first showing' at the dawn of the twenty-first-century to its current 'ubiquity'. We also investigated the question of whether it is possible to identify a 'learning curve' for these issues in China. It was in the period 2005–2007 that China's energy landscape took 'shape', in the context of record growth (up to 10 per cent) that gave rise to a greater demand for energy. This in turn required important reforms in the energy sector. Self-reliance could not longer be the catchcry, and it was acknowledged that energy security would increasingly become a function of oil imports. Securing steady supply became first and foremost among the Chinese government's priorities and 'energy supply' the key concept in driving energy policies in China. This can be seen in the strategy then adopted towards National Oil Companies (NOCs) that favoured their restructuring and overseas acquisitions. One consequence of the changing international energy landscape after 9/11 was that the predominant focus on energy supply began to be accompanied by a progressive change of vision regarding energy security. This led China to reconsider its strategic position in Central Asia and the Middle East, which are today China's main energy suppliers. Moreover, the growing demand – indeed the accelerated demand – for energy, accompanied by important episodes of energy shortages and blackouts in the years 2002–2003 and a deteriorating environmental situation, led the government to formulate a new approach to energy security in order to develop a comprehensive and effective strategy. As a result of being forced to grapple with these various issues, China began to engage in energy sector reform.

In the beginning, our principal aim was to identify the drivers of this reform in China, and to assess its progress and its constraints. We were interested in looking at how these drivers and constraints changed over time, as well as identifying the main factors for China's future energy choices. Over seven years later, our aim is now to at take stock of the ensuing situation. The 'shape' that developed in these years, based on a 'strategic approach' (Andrews-Speed *et al*. 2002), that gave pride of place to supply-side policies and procurement, revealed a signs of a rethink. Just when Chinese energy policy was 'shaping up', we noticed that the subject of security of supply was left almost untouched in academic debates. The supply side was the paramount priority, while the real situation was already showing a growing complexity due to an ever-increasing demand for energy. The point was that institutions had first to be consolidated, given that the existing framework was identified as a major component of China's energy insecurity. A new approach to energy security thus emerged under the new leadership of President Hu Jintao and Premier Wen Jiabao. From an institutional point of view, the creation of the *Energy Bureau* in 2003 was seen as an answer to the fragmented nature of energy policy management that had characterised China since the 1990s – spread across different institutions and lacking of the capacity to develop and implement coherent policies. It was accompanied by the creation of the *Leading Group on Energy* within the State Council, aimed at formulating the strategic direction of energy policy and coordinating various policies (Downs 2006: 16–19).

Energy strategy was also showing interesting developments in terms of contents, in particular by introducing an important focus to energy conservation and the sustainable use of energy resources. Increasing environmental issues and the problem of energy shortages brought about the development of a *Medium and Long-Term Energy Conservation Plan* in 2004, aimed at reducing China's energy intensity by 20 per cent by 2010. Considerations of energy conservation were accompanied by a recognition of the need to reduce the part of coal in energy choices, constrain the demand for fossil fuels and search for the 'right energy mix' that could help China reach the multiple objectives of its newly-shaped energy strategy. Three scenarios in particular were identified as possible approaches to China's energy choices: 'business-as-usual', exclusively focusing on energy supply; a scenario merging energy supply concerns with energy efficiency concerns; finally, making energy efficiency and environmental protection the top priority (Meidan *et al*. 2007: 55).

A number of pertinent questions were already being raised at that initial stage. First, this concerned policy choices, in particular in terms of their comprehensiveness and coherence. This was accompanied by an interest in policy tools, namely whether the government should adopt more economic and legal incentives or should still rely on administrative means. There also remained the critical question of implementation: lax enforcement in the context of a woolly legal and regulatory environment. Hence the need to develop better incentives for implementation. Responding to these challenges required the Chinese government to make energy a top priority and to develop a powerful agency for energy management, capable of coordinating the different policies dealing with energy security and ensuring coherence in the implementation of all related measures. The search for the 'right energy mix', including the search for alternatives to coal and oil, led the government to take new steps towards unleashing the existing potential. Beyond these main points, we also identified a clear allocation of responsibilities among the principal energy actors, the development of an appropriate legal system, clear access to information about rules and regulations for all energy actors, as well as access to relevant technologies. These represent the key elements constituting what we call the 'shaping' phase.

Some years later, faced with ambitious energy development plans and a situation characterised by growing complexity and poor achievements in energy production, the question of supply came to the fore in all its urgency, as did that of the rigidity of the 'shape'. The core question at the base of this new research was whether the processes and institutions that took shape with the development of energy policies in the 'mid-2000s' have allowed for effective control over the energy sector and achievement of the set targets. Surprisingly, what we have noticed is not only an evolution of processes and institutions, but also a significant change in the philosophy behind energy policies in China. We refer to this as a shift to the 'frame'. In other terms, the evolution of institutions and patterns of energy behaviour, resulting from a progressive learning process of the main energy policy-makers in China, were indicators of a shift from a rigid adherence to the shape to accepting the need for flexibility and 'realism' in the realm of energy. This configuration, which still seems to be ongoing, looks less constraining than the initial 'shape', and we found it therefore more appropriate to talk about 'frame'. The learning process implied by this shift from 'shape' to 'frame' resulted from a process of reconsidering policies and tools, realistically acknowledging the possibility of reforming energy management, and observing the main challenges to achieving the goals of energy policies. 'Flexibility' can be identified as the key word, as it has been particularly evidenced by the case of oil and of its management.

An illustration of this through a 'case study' based on oil in the next paragraphs will serve to better understand this shift from 'shape' to 'frame'. Oil has classically played the role of being a 'benchmark' of change in energy strategies. We decided to deal with this resource as a starting point for our considerations. We move on to a consideration of energy security from a number of perspectives – supply, international relations and energy conservation – as we feel that these are much key to describing the challenges currently being faced by China's energy strategies.

An approach to the general 'frame' of energy policies in China[1]

In 2007, the 'shape' that was taking place was based on the necessity to deal with the pressing problem of increasing energy demand in relation to economic growth. The energy shortages and blackouts of 2002–2003, together with the increasing importation of oil, were harbingers of fears that China would mobilise all of its financial resources to obtain raw materials, significantly impacting on oil markets. In the years between 2005 and 2008, oil prices increased sharply, further fuelling these fears that China would put all its energy into supplying its voracious appetite. However, after 2008 oil prices started to tumble and thus began a series of fluctuations over the following years. Could it possibly be that China, with its energy needs, was the 'price-maker', while the course of oil prices was clearly telling another story? These developments, and China's performances, cannot be properly understood without reference to the global environment.

As solid and steady as its growth may look (and probably because of this) China is not a unique example of a country to be very much concerned by its supply and access, the two structuring sides of the key issue here. In this context, it is worthwhile to remember that among the past emerging powers, and especially the UK and the USA, the rise of industry and the growth in wealth could be explained by the abundant supply of energy sources. There could have been no industrial revolution in England without coal. Equally, the USA would probably not have reached its level of wealth so quickly had it not been for the discovery of oil in the late nineteenth century. The emerging powers of the twentieth and twenty-first centuries are different, however. It may well be that China had initially benefitted from its existing reserves of coal and even oil, which the country exported until 1991. But the country's future and further rise, just as in the case of India, depends on the ability to fuel growth, and that is the 'supply' issue when it comes to using the domestic reserves. The question of supply thus became central. As for the issue of 'access', it is triggered by the deficit in the appropriate energy sources. Be it merely due to usage by the corresponding industries or to the consequence of an adjusting 'energy mix', China now also has to source its energy outside of its territory. Among the so-called 'BRIC' countries, only India is at least as much exposed to a shortage in energy supplies arising from the recent strong growth. However, energy intensity is on average strongest in China. Any threat to the supply of the extra energy necessary to fuel economic growth would be extremely detrimental to China's overall stability, both economic and political.

As with many political issues, Chinese authorities have always based their energy security policy on both an attempt to closely monitor the responses elaborated to address the various potential threats and to diversify the solutions and sources of supply. Outsiders have often been puzzled by this delicate balancing act, starting with the disappearance of the Ministry for Energy in 1993, the very year in which China became an importer of oil (crude and products), after years of being not only self-sufficient but also an exporter. It eventually became a net importer of crude in 1996.

The events of 1993 and 1996 would justify why these years should be considered as the initial cornerstones before the main episodes in the shaping of energy policies in China. Those were the years when China, turning into an importer of oil, could be described as 'joining the club' of oil importers, and it embarked on a trajectory that has not yet stopped, that is to say an ever-growing dependency on imported oil and non-domestic resources. At that time, these events were also giving very early signs of things to come in terms of the growing Chinese economy, where oil dependency was increasing with the steady pace of economic growth. Although China's accumulated wealth and volume of international trade were at the time far from being up to the standards of the industrialised and largest trading nations and its GDP still ranked very low in the global charts, the country started to face issues which Western Europe and the USA had been confronted very early on. Those issues were twofold: on the one hand, there was a need to build a coherent policy ensuring energy supply; on the other hand, a choice had to be made among the best tools and agencies designed at implementing this policy. These issues are precisely the ones we have identified as the 'shapers' of the future of Chinese energy policy, from a (relatively) easily planned allocation of resources and duties between and within state-controlled companies, to the need to make sure that future growth would not be hindered by an inadequate domestic supply. Beyond the concern for quantity, a focus on quality also started to emerge. The issue of ensuring the proper and smooth domestic supply gradually gave way to a concern about the appropriate mix of energies, and thus about the way to access them, both from a geopolitical and technological standpoint.

Oil and oil 'decision-makers' as the main evidence of a 'learning curve' taking shape

A series of drastic changes followed, which in turn shaped not only China's energy policies but also its politics. This aspect is related in the first place to the excessive importance of the 'oil issue' in energy policies and of a group of people closely involved in shaping oil policy. Certain players thus rose to prominence in the early years of the century, after it became clear that the new 'shape' within which the energy policies were being designed was somehow becoming stabilised, in particular through the division of labour among the main national oil companies (NOCs). Indeed, after the demise of the Ministry of Energy in 1993, three companies (formerly ministries) were given specific tasks in energy management: the China National Petroleum Corporation (CNPC) covering the role of exploration and partially refinery; the China Petroleum and Chemical Corporation (SINOPEC) being responsible of refinery; and finally the China National Offshore Oil Corporation (CNOOC) working on offshore operations.

The people in charge were either already heads of the three NOCs, with a status equivalent to that of cabinet members and those within the top echelon of the Party, or very influential academics or members of the powerful National Development and Reform Commission (NDRC), the country's main actor for economic planning. Such profiles can be best illustrated by Fu Chengyu, a long-time head of the CNOOC, before he took over as number one in the NOC and CNPC. Among his remarkable achievements as the head of the CNOOC, and in line with Chinese strategy to have overseas investments guarantee security supply, Fu Chengyu acquired five 'blocks' in Indonesia from the Spanish oil company Repsol in 2002, turning the company within a year into the Chinese largest offshore operator. During the years of his leadership in the CNOOC, Fu Chengyu also had the CNOOC purchase a 5.3 per cent interest in one of the first liquefied natural gas projects in China, in Guangdong province. Many other examples, including his failed bid to acquire UNOCAL (see below) give an idea of the stature of this quasi statesman. Others, such as Ma Fucai and Jiang Jiemin, the latter being the son of former PRC leader, Jiang Zemin, are also indicative of this 'oil team' which sometimes had also been dubbed the 'oil gang'.

Just as it is important to identify the years 1993 and 1996 as being crucial for China's energy policies, and as an evidence of the important role played by Fu Chengyu, for example, the year 2005 should also be seen as a key date. Between 1993 and that year, the profile of China's oil imports and its overall consumption changed quite drastically, as it shown in Table 1.1.

Nevertheless, 2005 has to be remembered above all as the year when the CNOOC failed to take over the US company UNOCAL. Four years after China joined the World Trade Organisation (WTO), which was supposed to open the doors to multilateral investments, the CNOOC made an 'all-cash' bid (US$8.5 bn) to buy the US company, topping an earlier bid made by the US Chevron. Finally, after the decision to put this to a shareholders' vote at UNOCAL, the CNOOC announced the withdrawal of its bid, pointing to existing political tensions in the USA. The threat of a veto by the US Senate, concerned about the sale of a domestic company to one owned by the Chinese government, made the CNOOC back down from its attempt. From these events, questions emerged as to the role played by the Chinese government, in terms of its influence over the bid as well as regarding the decision to withdraw. Questions also emerged about possible deals made between the USA and China in this process.

The UNOCAL event anyway marked an important turnaround in the story of Chinese NOCs, displaying the increasing power of the Chinese national

Table 1.1 China oil imports/exports and oil consumption (1990–2012)

	1990	1993	1996	2000	2005	2012
Imports of crude oil(ktoe)	2,923	15,671	22,617	70,265	126,817	271,027
Imports of oil products	3,568	18,377	19,976	24,963	41,494	52,699
Exports of crude oil	−23,990	−19,435	−20,403	−10,306	−80,670	−2,432
Exports of oil products	−6,654	−4,551	−5,493	−10,237	−167,220	−28,744
Total consumption of oil products	81,958	107,043	134,992	177,790	267,239	420,670

Source: IEA Statistics.

companies and above all their potential to enter the oil market of the major world players, such as the USA. It also raised some questions as to a hypothetical process of learning of China's energy decision-makers. In a strategy aiming to ensure energy security, the main focus of overseas acquisition is not properly to obtain a company's share but rather its assets: its oilfields, resources and technologies. Doubts emerged as to whether the CNOOC engaged in due diligence when attempting to buy UNOCAL's shares, but in hindsight one could it would appear that the failed attempt was a fortunate event for the Chinese company and as possible signs of a reconsideration of its energy investments and a more cautious attitude.

In actual fact, and against a background of mixed feelings (insecurity of the Chinese company, a newly displayed 'arrogance', as interpreted by the USA) and a test of American determination to protect its own backyard, the CNOOC took the opportunity of UNOCAL to appear as a potential buyer. Above all, this attempt and the acquisitions made abroad showed an important element in energy behaviour, potentially indicating the presence of a learning process. The CNOOC, whose responsibility had been defined following the new distribution of roles in 1993 as being in charge of oil digging, drilling, exploring and prospecting national offshore sites, was ready to blur the lines and play a new role in the energy 'division of labour'. Under Fu Chengyu's administration, the CNOOC started to make investments that would have hardly been permissible within the previously identified 'shape' of the energy sector. Bypassing the established competences for each company was indeed an indicator of increasing competitiveness among the NOCs. Instead of respecting the specific tasks attributed throughout the whole 'oil chain' (from exploration and prospection to refining and distribution), the 'Chinese majors' started to compete for growth and action like any other energy company. This fact also showed that Chinese energy security had shifted from being a purely government (and 'governance') matter to being one of the main stages for political choices, and personal career building.

Whatever the logics behind the attempt to buy the North American company and whatever the explanations for its failure, the UNOCAL event had also several consequences, both nationally and internationally. For outsiders, the image of an 'oil-hungry' China took hold and probably distorted their vision of the Chinese energy mix. The trend of 10 per cent GDP growth made commentators think that it was likely to have been mainly fuelled by oil. In the light of rising oil prices registered in those years, the main theory adopted by scholars and commentators was indeed that China played the role of 'price-maker' in world markets, heavily impacting on the increase in oil prices. However, after 2008, oil prices started to fluctuate, invalidating the theory of China's supposedly heavy influence. In reality, although oil is an important energy source for China, it is not the most widely used one.

This focus on oil issues for some of those analysing Chinese energy policies meant that they started not to see 'the forest for the trees'. Although China is heavily reliant on oil imports, which now amounts to 56 per cent of its total oil consumption, it still has alternatives to oil. In fact, Chinese policies to acquiring shares and signing supply contracts abroad are not only a means for reaching out to feed a supposedly 'resource-poor country'. Its policies are also related to exploring new technologies (as the case of the acquisition of Nexen shows, see later), improving infrastructure in energy supply and distribution, diversifying the reliance on different energy sources, and improving energy efficiency, while looking at the climate and economic impacts of its energy choices. Indeed, if there is a need to talk about a potential threat linked to Chinese energy consumption, its focus should not be on the impact on the availability of oil supply, but more on the impact it has on climate change, as China became in 2008 the main producer of carbon emissions.

In China, the failed attempt certainly led to a totally different approach to securing energy supply. The lesson that was learnt was indeed the start of a diversification of strategies for acquiring oil assets (Meidan 2007), leading to smaller, less visible purchases, as well as a greater involvement in cooperation contracts. Moreover, it led to a re-thinking of the cooperation between the various bodies involved in shaping energy policy, still characterised at that time by a rigid division of labour as established after 1993. In particular, the problems identified in the area of energy governance brought about an evolution of the two institutions established in 2003, the Bureau of Energy and the Leading Group on Energy. In 2008, claims that a more powerful energy agency – another Ministry of Energy – had been established were circulating among energy commentators, but this never materialised. Instead, the existing Bureau was replaced by the National Energy Agency (NEA), while the Leading Group became the National Energy Commission (NEC). In particular, the former kept the functions of the previous Bureau, while the NEA took on both the tasks of the Leading Group and of the former Department for Energy Efficiency within the NDRC. With this reorganisation, the two bodies can be seen as having been involved in shaping the main directions of energy policies and regulations by sharing their duties and gathering the important components of energy efficiency and energy transition. However, notwithstanding this reorganisation, and as illustrated in Fig. 1.1, the entities involved in the decision-making process mean that it still remains quite intricate.

The NEA's coordination capacities are still largely limited by a separation of tasks which gives responsibility for setting energy prices to another NDRC department, that of Prices. Moreover, the supervision of progress in energy efficiency is in the hands of another NDRC department, that of Environmental Protection and Resources Utilisation. Beyond these divisions of competences within the NDRC, other institutions still keep a firm hand on energy questions. Whilst the three big NOCs have been guaranteed more latitude in their investment choices and the possibility of working in competitive conditions, they remain firmly within the hands of the State-owned Assets Supervision and Administration (SASAC). This institution owns the majority of their assets, controls their strategies, and supervises their reforms and

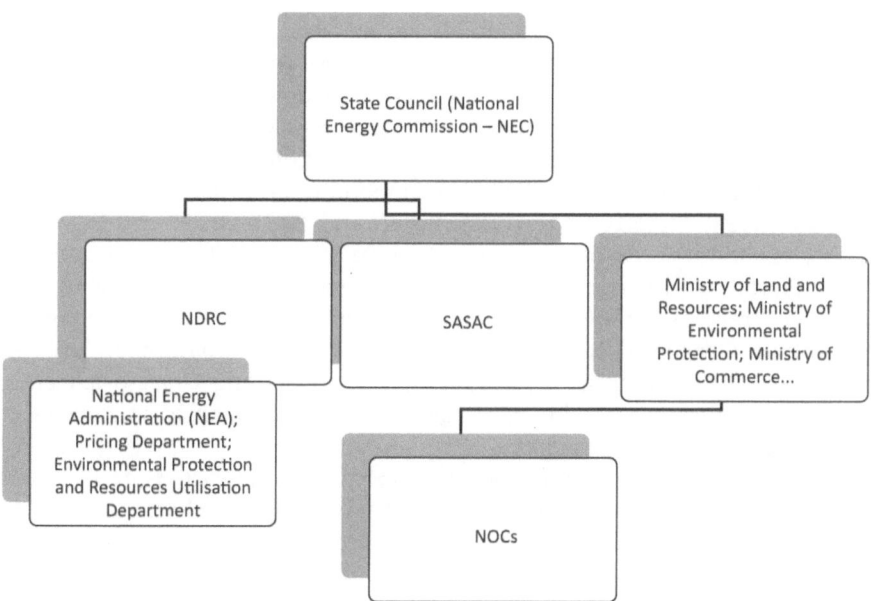

Figure 1.1 China's energy administration

restructuration, in addition to being in charge of the appointment of board members. The NOC's activities are also controlled by the State Council and its NEC, as well as by the NDRC and by different Ministries.

This configuration of energy governance, which did not really offer a solution to the fragmentation of energy decision-making in China, makes evident two important aspects that can be attributed to the concept of 'framing'. Firstly, the idea of having a powerful actor that concentrates the competences over all aspects of energy policy and reduces conflict among the main energy actors – a new Ministry of Energy – was a solution unlikely to prevail. The 'feudal' attitude of ministries, agencies and energy companies showed that no one was ready to give up any of their powers. The NEA, when it was created, was supposed to tighten coordination and introduce a more clear-cut hierarchy. However, history tells us that NEA did not really fulfil this role, for reasons of insufficient staffing and gathering of competences. For these reasons, it is much better to talk about 'framing' rather than 'shaping'.

A second, inter-connected, aspect is the flexibility of this configuration, further justifying the use of a 'frame' approach. Not only the configuration acknowledged the patrimonial division of competences among these actors, but it also revealed that it could function in a relatively flexible way in an energy policy context still characterised by a mix of planning and the market. The NOCs are a clear case in point. In order to increase their supply capacities and improve their performance, the government allowed for the introduction of competition between these companies as well as a certain room for manoeuvre concerning investments in foreign markets. As result of vertical integration decisions, companies were eventually transformed into groups, obtaining the possibility to operate both upstream and downstream (Bo 2009: 15). However, as energy security concerns at the domestic level also coincided with the need to guarantee an equal access to the population, price control by NOCs shows how the administration still plays an important role in energy management. The case of gas imports from Turkmenistan and the losses it entails for the CNPC, as shown in Chapter 3, is one example of this flexible use of administration tools. Another aspect showing learning and flexibility in the shift from 'shape' to 'frame' is represented by the opportunities given to NOCs to control instruments of financial security, particularly trading and arbitrage. At present, the largest portion of their asset acquisition is not feeding Chinese refineries, but world markets, which also once again shows that China is not 'obsessed' by the need to ensure oil supplies.

It is not the ambition of this book to elaborate on the various iterations of Chinese domestic politics or on the changes in internal debates and power structures. These points were extensively covered in our earlier work, since energy governance was in 'shaping' mode at that time. We have merely summarised some key points concerning current energy governance in China, as a result of these reforms, and show those aspects that can delineate a shift from 'shape' to 'frame'. In this regard, energy policies are an interesting key to better understanding the major trends and influences within the world of Chinese politics, as energy security had become a more complex issue, giving rise to a 'liberal' approach to sourcing oil energy resources.

Parallel changes in China's domestic energy and foreign energy policy mixes

As previously mentioned, China's energy mix has for the years been a quite peculiar one within the group of major economies. China is a very atypical case in terms of the global energy situation, which is clearly evidenced by Fig. 1.2, representing China's primary energy by fuel type over the years.

The most outstanding peculiarity is surely the place of gas in China's overall energy mix. World data show that it has a 24 per cent reliance on gas for its overall supply, which is higher even than that of the USA (where shale gas partially distorts the picture), and various European countries, which reach 30 per cent in terms of gas usage (except in France, where gas accounts for only 16 per cent) China is lagging far behind:[2] gas usage in China represents just 5 per cent of its current consumption. Another specificity is represented by the opposite roles of oil and coal. In China oil only accounts for 18 per cent of China's energy supply as compared to 67 per cent for coal. China's coal consumption is too high to have any significant impact on the global ratio. The most relevant economies use coal for less than 20 per cent (the USA included). The global picture shows a rate of 30 per cent, while in the ratio of non-OECD countries coal represents 38 per cent, a figure that is

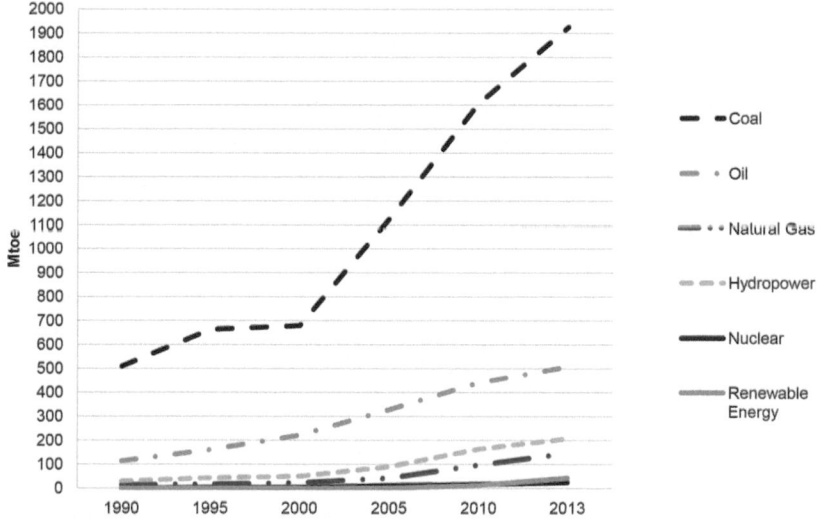

Figure 1.2 China primary energy consumption by fuel type
Source: Data retrieved from BP Statistical Review of World Energy (2014).

justified by the weight of Chinese coal consumption in the world's economy. Conversely, oil accounts for an average of 30 per cent in energy consumption for most countries outside China. The only significant exception to this 'normal' mix is France, where nuclear energy accounts for almost 50 per cent of energy consumption (including the transportation sector) as it covers more than 70 per cent of the country's power generation.[3]

If we look more closely at the place of coal and oil in the energy mix, the policies adopted by the Chinese government show a continuity with the early years of the millennium and already point towards elements of change. These elements of change can be observed for instance in the interpretations of Chinese scholars. At the beginning of our research within the framework of Asia Centre activities, the impression we had from our exchanges concerning the approach to oil was that of a 'hungry and anxious giant', somehow confirming the analysis of foreign commentators depicting China as a 'hungry dragon'. Eight years later, it seems that Chinese decision-makers, while not totally addicted to the 'market', are increasingly 'relaxed' and have adopted (at least implicitly) the attitude whereby 'there will always be oil for China as long as China can pay for it'.[4] These observations reveal an important change of attitude and a learning curve both at a domestic and an international level, as evidenced by oil policies, among others.

Searching for the 'right mix' or just 'mixing'?

In terms of domestic policies, the current debate is mainly focused on discussing about the proper and future energy mix, where there is evidence of a change of paradigm and the emergence of different views, according to which coal and oil resources are still not elements which, in a changing environment, would constitute fixed 'swing factors' for the energy security landscape in China. Those would rather be gas (and mostly shale gas), nuclear energy and renewable energy. The consideration of several resources – with an important role for gas – shows that there is no unique desirable mix, but potentially several mixes or at least a certain fluidity in the definition of the mix. In other terms, while in the past it was possible to observe a certain rigidity in desirable energy mixes, as evidenced in the five-year plans, recent trends show how much the mix could be flexible, with an important, albeit diminishing, role still to be played by coal (though reducing) and the progressive entrance of new 'actors', namely gas, among the different developments in their respective sectors.

The coal conundrum

There are many simple explanations for the current predominance of coal in the Chinese energy landscape, as well as many probably also pointing towards a decrease in this prevalence. China still holds huge reserves of coal reserves, estimated at around 200 to 300 years of supply.[5] For a long time it will remain the yardstick by which other sources of energy will be assessed, since, contrary to oil, it does not carry the so-called 'curse'. An abundance of oil usually occurs at times of political instability and geopolitical conflicts, but this is not the case for coal. However, the predominance of coal is supposed to decrease for at least two reasons. The first is the need to upgrade the varieties of coal used in China, in order to lower pollution levels, and the second being the need to upgrade the logistics of coal, which would clearly end up being very costly. These two factors will make the edge enjoyed by coal in terms of price competitiveness less and less obvious and help promote alternative usages of coal (as will be shown in Chapter 2, dedicated to coal and its various forms).

Although various school of thought used to consider coal not as relevant to the issue of energy security, the predicted evolution in the approach to coal that China will adopt, as well as the commitments it may make vis-à-vis climate talks, will tend to make coal an energy source more and more comparable to others, including in the way it is managed. As extensively demonstrated in Chapter 2, 'coal governance' remains weak. Notwithstanding the attempts to rationalise coal production, there is still an implicit tendency to let things go unmanaged as they used to be, with a split between large coal companies based in Northern China and other smaller, less regulated producers scattered throughout Shanxi province and southwest China. In the future, coal governance could become much more like the systems prevailing for other energy sources, so would most likely be more developed in an integrated manner, while price policies and infrastructure building could become more coherent.

Oil & co.: indicators of policy changes

As mentioned above, oil represents only 18 per cent of primary energy consumption in China, while coal remains the main resource, both on account of its broad availability within China and the weight of its industry. Equally, power generation can be said to be 'addicted' to coal, which remains easily accessible in spite of the many obstacles, namely in its transportation, as evidenced in Chapter 2. However, when tracking energy policy changes in China as well as the various historical changes of tack, we see that oil remains one of the best indicators. China's domestic oil demand has reached 9.8 million barrels per day (mbd), almost double its 2003 level, contributing roughly 60 per cent of global energy demand growth every year throughout this period. There is obviously no question about the future increase in quantities of oil that China will need to import or about the impact of its oil needs on the global routes transporting hydrocarbons. Yet the Chinese oil landscape has dramatically changed following two key domestic events, and then a series of other episodes that had an impact on China's 'oil patch'.[6] The most interesting feature is that these two key events probably also impacted on the future developments of Chinese oil companies abroad, and yet their most important effects are still to come.

The first and earlier of these two events was the removal of several NOC heads in 2013 and other changes at their helm. These changes were in a sense a signal that the industry management was still firmly in the hands of the central government. The second big and more important decision was the surprising dismissal of Zhou Yongkang, China's 'oil tycoon'. Between the middle and the end of 2014, Zhou gradually lost all his official positions, making the whole industry somewhat of an 'orphan', shaken but not totally transformed. Zhou Yongkang was of course of the main people responsible for the great expansion of NOCs. During the rule of President Hu Jintao and Premier Wen Jiabao, Zhou Yongkang had for many years been in charge of the Ministry of Land and Resources. He then served as the Head of National Security, before the handover to the new government 'tandem' formed by President Xi Jinping and Prime Minister Li Keqiang – at the top of the political elite in China. The event that shook the whole oil industry without entirely turning it upside down has still perhaps to show its full effect. It is too early to tell, but questions are emerging with regard to the lack of a subsequent change of all the heads of NOCs, whose fortunes are related to that of Zhou Yongkang.

Another series of episodes impacting on China's oil policies, among others, relate to China's role in the game of international relations between regions and countries. These events at once show a capacity to display an excellent sense of timing and an ability, at best, to seize opportunities, as well as too much 'haste' when it comes to certain energy investments – mistakes from which China is learning. For example, the gas deal signed with Russia in May 2014, which had been under discussion for almost ten years was concluded at a time when Russia most needed to secure some outlets for its production. The budding Ukrainian crisis was alienating most Western countries from Russia, but turned out in the end to be advantageous for China, enabling it to secure the best possible deal with Russia. The deal made with Russia was also prepared by an 'encircling' of Russia, a strategy that had been made possible thanks to newly appointed President Xi's first trip abroad, in March 2014. As shown in Chapter 7, President Xi strategically chose Central Asia as his first destination, one of the upshots being to secure strong gas alliances with Turkmenistan.

In contrast to this brilliant strategy, more related to gas resources, other forays show certain shortcomings with regard to China's game and positioning in the oil market. For example, China's acquisition of the Canadian energy company Nexen in 2013 can be seen as the product of a strategy that was not well thought through. The acquisition of Nexen's assets cost CNOOC $15.1 billion, which found this opportunity to 'take its revenge' for when it had the missed its chance to buy UNOCAL. At the time, China had allocated considerable sums to encourage NOCs' overseas investments. The decision to buy Nexen was particularly justified, as the company held important oil sands assets, a significant resource for China's diversification strategy. However, the decline in oil prices made these products less attractive and ended up turning this huge investment into a 'headache'. CNOOC now has to cope with the typical problems of this kind of resource, such as high extraction costs, operational challenges and volatile bitumen prices.

Another example of 'failure' and subsequent learning from the market has been the purchase of Addax Petroleum, originally belonging to the Addax and Oryx Group founded by a French-Swiss businessman and operating in the African continent. In 2009, Sinopec bought Addax Petroleum, transforming the company in 2011 into Sinopec-Addax Petroleum Foundation. The mistake made by the Chinese company concerned a clause in the deal, which included the possibility to keep the previous directors. Since Sinopec had bought Addax Petroleum, the former executives founded in 2010 a replica rival company, Oryx Petroleum.

These events have constituted important learning opportunities for Chinese oil companies, including the events in Iraq, as we will show in the next paragraph. During the discussions with Chinese experts on energy policies,[7] our hypothesis of decreasing energy concerns for China was confirmed. This point is intimately related to important lessons learned in these years thanks to the overseas experiences of big energy companies. These forays have been important occasions to learn more about energy markets: NOCs are now better aware of how they work and for this reason have less cause for concern about the chance of disruption to supply. We were also confirmed in our view that the learning curve is also a function of the complexity and the size of Chinese economy, which is now reaching a level that puts the country in a better position to deal with risks, especially when these are market based.

The Middle East: learning effects and current debates

In the Middle East, against a general background of apparent oversupply and global economic slowdown, it was not a good time to invest in Iraqi oil exploitation, although this seemed at first sight to be quite a smart and bold move. Finally, while the large NOCs such as CNPC seized E&P contracts ('exploration and prospection/production'), the price paid both for the contracts and for the costs of developing these ended up being too high compared with the new benchmarks for oil prices.

On a related topic, it would appear that China is not involved in any significant way in the fixing of prices. The fluctuation in oil prices, instead of registering a steady increase since 2008, is already a sign that China, notwithstanding its position as the top importer of oil in the world, does not determine the market. The mistakes made in investing in Iraq are further proof of this. Instead of manipulating the market to correct these mistakes, for instance by heavily speculating on oil prices and very often playing with commodity derivatives, which represent a huge multiple of 'physical transactions' (as major oil actors would have done), Chinese energy companies rather tolerated the losses registered. The same can be said of the issue of securing oil routes, where China still leaves others to 'do the job', namely US ships, regardless of the potential new involvement of the Chinese fleet. As important as oil is and will remain for China, the question of the country's role in oil flows has not really evolved in the last 10 years. Although it is a key market player as a trader, and it is certainly a heavyweight when it comes to purchasing oil assets or winning contracts, it is not in a proper position to compete on a level playing field with the USA and other Western majors.

Concerning these points, there is a debate in China about the 'geopolitics of oil', which obviously translates into a difficulty to assess the future direction of the country's involvement in the Middle East, as we shall see in Chapter 6. The presence of this debate also shows that there is a gradual reconsideration of existing energy paradigms. Whereas there was previously a single common objective, consideration is now being given to alternatives. We touch here upon the borderline between China's internal politics and China's role in the 'multipolar world'. There are plenty of reasons for involving China in global issues; indeed, China itself is displaying an unwavering willingness to help in peacekeeping operations as well as a real capacity to be active both when unquestionable issues are at stake, or when its nationals are threatened. For instance, China committed its maritime forces in anti-piracy actions in the Gulf of Aden (Somalia) in 2008. On this occasion not only did China play a significant role, in what was an important opportunity to learn how to deal with this type of operation, but it also showed a new commitment to shared governance, by understanding the benefits it can bring, especially in the management of raw materials, and perhaps thanks to a 'market' culture that is progressively taking root. In Libya, during the fall of the Khadafi regime, China registered very important financial losses after years of investments.

But it also particularly distinguished itself for its engagement in a rescue operation of its citizens from the threatened zones. These are signs that China is going to participate more and more in other issues beyond raw materials.

Still Chinese behaviour vis-à-vis the Middle East is not yet commensurate with its growing involvement. There is one school of thought (Meidan 2007) that would rather support giving benefits to oil-producing countries in exchange for secured oil supply. Such an attitude would warrant a hedge against the 'Western monopoly' in Saudi Arabia, for example. For these proponents, Chinese action (including at the level of United Nations) could even become an effective leverage for exporting countries on the international scene. As a complement to this policy – if not its converse – the investments in oil fields and resources abroad have also been seen as a way to secure easier access. However, and while fears have been raised throughout the world about the growing and supposed threat of a new player in a strategic sector, the outcome of this foray remains quite subdued, with two-thirds of the investment not reaching break-even (even before the drop in the oil price) and only a tenth of the extracted oil being directly imported into China.

Gas and its role

Gas is probably the main alternative to the overarching dominance of coal in the Chinese energy mix. As a matter of fact, the potential is huge, greater even than for oil, which will have to keep pace with the development of transportation and hence 'at best' maintain its share in the mix. The 12th Five-Year Plan (2011–2015) certainly provides for an important increase in gas usage, and Chinese diplomacy has definitely been pro-active in this respect. However, in order to reach the consumption objective of 230–260 billion cubic metres/year by 2015, China would have to import at least 35 to 40 per cent of this amount. As we shall see in Chapter 3, the gas development objectives, as well as the other objectives included in the Plan, pose important challenges to Chinese energy policies. This raises three main questions:

- Will this ambitious plan, whose aim is to bring China to today's level of most developed countries, succeed? The projected mix would make China's energy mix and the weight of gas in this mix close to most developed parts of the world, but the road ahead will be long and somewhat arduous.
- How will this plan involve urban distribution of gas, beyond the very classical involvement of oil majors?
- What will be the real share of shale gas?

As for this new resource, and in spite of the water shortages in China, the potential of shale gas potential has been loudly proclaimed in China. Before the announcement of the Russian deal, one of the main sources of an increase in gas output was supposed to come from this product. However, the numbers released in 2013 showed poor results in the exploitation of this resource, what made the target, as projected by Sinopec, CNPC, Yanchang and other main actors in shale gas exploration, quite unrealistic. Time will tell whether, as has been often mentioned, 'delivering a message' is of greater importance than actually executing an ambitious plan.

Nevertheless, when added up, these elements concerning the place of gas clearly point out to a significant, albeit gradual, shift in gas output and imports for China. The underlying goal is extremely critical: apart from the development of nuclear industry, the only chance for China to solve the 'coal conundrum' and achieve the stated goal of lowering coal dependency to under 50 per cent by 2035 lies in the aggressive development of gas resources.

Nuclear energy and renewables

Concerning the so-called 'fifth renewable energy', i.e. nuclear power generation, China has once again put itself in a very peculiar position. As we will see in Chapter 4, China has clearly reformulated its ambitions, even after Fukushima, which had triggered a profound slowdown, and in some cases a complete standstill, in nuclear projects all over the world. China has clearly reformulated its ambitions. Yet after a short period of supposed questioning, new goals have clearly been set.

The case for nuclear is interesting also because in addition to being one of the very few countries still adhering to a massive expansion of nuclear energy, China will also launch an 'indigenisation' of the nuclear industry and become one of the major global competitors in this field. In contrast to other energy industries, this one will not only be 'China-centred' but will also become one of the most visible developments of energy-related technologies to emerge from China.[8]

Speaking more specifically about genuine 'renewables', and even if hydropower is included in the mix, we can say that they are bound to represent 9.5 per cent of China's power generation (see Cassisa and Denjean, this volume, Chapter 4). As Chapter 4 will show, this is probably the most intriguing feature of the Chinese energy landscape. It can be seen as representing the 'flip side' of the predominance of coal. China has long promoted not only the use and development, but also the manufacture of far-reaching, state-of-the-art windmills and solar panels. It has, however, encountered strong opposition in trade talks about this very strong push. Nevertheless, just as the predominance of coal might have to be revisited, this 'virtuous' side of China's energy landscape still faces many hurdles, as it will be particularly shown in Chapter 9. China's geography as well as its positioning vis-à-vis the benefits deriving from the 'Kyoto protocol'[9] (initial version) would logically lead the country to appear deliberately as a promoter of renewable energy. However, it is more than probable that the switch in energy mix, together with the 'new norm' which is producing the new 'frame' of China's energy mix and future, might rely more on other factors than the mere (and marginal) one of renewables.

Energy efficiency as the 'potential saviour' of China's energy security

As a matter of fact and in order to conclude this review, one of the most important developments in terms of energy security for China lies in energy efficiency. China's development has been extremely energy-intensive, as both Chapter 10, focusing on industry, and Chapter 11, focusing on cities, will show. Policies and actions have been developed both for the industrial sector and in urban development, but there are still doubts about their ability to contribute to energy security.

In relation to the topic of oil, it is interesting to observe how determined the Chinese authorities can actually be now in terms of curbing energy inefficiencies and overconsumption. The example of retail gasoline prices after the fall in the oil price is an interesting case in point. Contrary to what happened in the past, when taxes were still relatively low, the Chinese authorities have taken advantage of this drop to maintain a relatively high retail price and hence curb any trend to overconsume. In November and December 2014, as crude oil prices were dropping (by about US$70 per ton, compared to early 2014, i.e. about US$10 per bbl), the consumption tax had been increased by an equivalent amount, thus limiting the net change to either nil (in November 2014) or a maximum of 200 RMB per ton, i.e. US$3 per bbl. As for retail prices, while in the USA in late 2014 they were back to late 2008 levels (US$2.5 per gallon), in China, they had doubled compared with the same date (US$4 per gallon) (Wang 2015).

Book overview

This work aims to present a concrete picture of the prospects for energy security in China. It begins with an overview of the various factors in play showing that China's energy problems have changed over the last decade. These changes affect three aspects of energy security, which we analyse separately. In particular, we asked the contributors to address three basic issues: the place of their topic with respect to China's overall energy security; the principal policy tools and strategies; and the observations that can be can be made about the implementation of policies or application of strategies. Their brief was always to bear in mind the broader perspective of energy security. As the topics are various and touch on different dimensions, we allowed the authors a certain amount of freedom to interpret the questions and identify the points they consider to be crucial for China's energy security. The contributors' findings are presented in the conclusion of this collective work, underpinning the idea that in order to understand energy security in China a semantic adjustment from 'shaping' to 'framing' is required.

A first part is constituted by the domestic logics and interaction between the key sources supplying energy at home ('the supply side'). In particular, Chapter 2 by Liu Zhi and Wang Zhe, shows how predominant the role of coal is in China's energy security, as well as the need for a significant restructuring of its governance in order to promote new uses and guarantee overall energy security. Conflicts of interest and administrative interferences, in particular, are hindering the development of this sector. In Chapter 3, Giulia C. Romano, Yin Na and Zhang Xuanxuan, introduce the place of gas in China's energy security by presenting the ambitious targets of the Chinese government as well as the challenges that the country has to face in promoting gas as an important factor for energy security and clean energy production. Again, governance in this sector proves to be the crucial factor hindering the development of gas as an important component of China's energy security. Still in the realm of 'clean energy', Chapter 4 focuses on the place of renewable energy and the nuclear sector in China's energy security. Cyril Cassisa and Benjamin Denjean, engineers specialised in these sectors, introduce the state of technology development in China, its evolution and the shortcomings limiting the full development of these types of energy.

Turning to the international dimension, the various contributions also examine the key relations structuring China's energy policies worldwide. In Chapter 5, François Bafoil focuses on conflicts in the South China Sea regarding the possible rich deposits of energy resources there. Given the necessity to contain China's hegemonic power and to find a balance among the different actors interested in these resources, he discusses the search for a lasting compromise between the interested parties and those who gravitate in the area (namely the USA). Chapter 6 brings us to the Middle East, where Michal Meidan analyses the evolution of China's attitude to this part of the world. She calls for its progressive engagement to a stability of its energy provision, at the same time showing the country's reticence to fill the void left by the USA. Conversely, a more active role can be found in Central Asia, where the rich energy resources and fertile trade relations have seen China enter into the historical 'backyard' of Russia. Chapter 7, by Vanessa Boas and Olga A. Spaiser, takes the reader through the evolution of China's trade and energy relations with individual Central Asian countries, clearly showing China's progressive supplanting of Russia, in relation to the increasing energy demand of the 'dragon'. Finally, in Chapter 8, Yu Kaho and Zhou Yunheng analyse China's presence in African countries, by tracking Chinese energy investments and their significance to China's energy security. The authors argue that energy cooperation with African countries still needs improvements, namely in terms of China's relationship with African countries within the overall international energy system.

Our overview ends with a look to the future and the 'sustainability' of the energy situation, which also leads us to consider the role of energy transition, energy conservation in the industrial sector and, lastly, energy conservation and energy efficiency in urban development. Chapter 9, in particular, focuses on the topic of energy decentralisation as a way to contribute to energy security. In this chapter, Marie-Hélène Schwoob observes, from the perspective of energy governance, how the control of the main national energy companies constitutes important hurdles for the development of decentralised energy sources. This chapter also offers insight into the difficulties for the development of renewable energy from the standpoint of reaching a transition to cleaner energy, thus echoing aspects of the discussion in Chapter 4. Energy saving, another component of energy transition, constitutes also an important step for China's energy security strategies. In Chapter 10, Patrick Schroeder offers an overview of the policies and efforts undertaken by the Chinese government, identifying the achievements and the challenges that remain. Based on the observations made during the period when best practices projects were conducted for the industrial sector, he identifies the problems and the needs currently affecting industry stakeholders, limiting the possibility to pursue effective energy efficiency policy. Shortcomings in the application of energy efficiency policies are also identified in the eleventh and final chapter, focusing on the place of cities in China's energy security. For Giulia C. Romano, cities constitute the 'micro-dimension' of energy security in China, a dimensions that cannot be neglected in the context of a concern for the overall coherence of energy strategies. The author argues that if cities are not taken into account in the energy security equation, strategies developed at central level can prove futile, as current local performances suggest an important neglect of energy efficiency and energy conservation in urban development. As cities represent 75 per cent of China's current total energy consumption (and 83 per cent by 2030), 'classical' energy management policies need to be coupled with strategies at the urban level, namely in the development of new spaces and the renovation of existing ones.

Notes

1 Most of this information comes from Asia Centre's research on energy in China in the various years of our research project, starting from 2005. As a personal focus of J.F. Di Meglio's analyses, the author had several exchanges with experts, including representatives of both international and Chinese companies, Chinese ministries and NDRC representatives, scholars, think tanks and representatives from embassies.
2 OECD. Available at: www.oecd.org/china/publicationsdocuments/reports/2014.
3 Ibid.
4 Personal conversation with CICIR, Beijing (2014).
5 OECD report.
6 This is the traditional name by which the group of people ruling the oil industry are called in North America, Canada and the USA. The word could also apply to similar groups in China.
7 Personal conversation with CICIR, Beijing (2014).
8 Forays in the hydropower have indeed been done, but without really transforming China's role in this type of engineering at this juncture, especially when one thinks of South East Asia (Laos) where China has also encountered some disappointing situations.

9 As a 'developing country', China was bound by any constraints in the initial version of the 'protocol', despite being a huge polluter. However, subsequent versions of the protocols regulating emissions and their trade were to affect this position.

Bibliography

Andrews-Speed, P., Liao, X., and Dannreuther, R. (2002) 'The Strategic Implications of China's Energy Needs.' Adelphi Paper No. 346. London: International Institute for Strategic Studies.

Downs, E.S. (2006) 'The Energy Security Series: China.' The Brookings Foreign Policy Studies. December.

Meidan, M. (2007) *Shaping China's Energy Security – The Inside Perspective.* Paris: Asia Centre.

Wang, T. (2015) 'How will lower oil prices impact China growth?' UBS economic briefs (21 January).

2 Coal resources and coal utilisation in China

Liu Zhi and Wang Zhe

Introduction: the place of coal in China's energy security and its challenges

As the predominant energy source of China, coal accounts for 82 per cent of the country's primary energy output and 72 per cent of its consumption. Energy security in China thus currently goes hand in hand with coal provision and consumption, making the country the world's largest producer and consumer. According to the Chinese Ministry of Land and Resources (MLR 2013), China's coal reserves represented 1377.89 billion tons at the end of 2011, while the remaining technologically recoverable reserves of oil and natural gas were 3.24 billion tons and 4 trillion cubic metres respectively. Coal reserves are plentiful yet unevenly distributed, and are primarily based in northern and northwestern China. In terms of geological distribution in administrative provinces (Cui 2011), coal resources in Xinjiang Province and Inner Mongolia Province take up 66 per cent of the total reserves, while five other provinces, Shanxi, Shaanxi, Guizhou, Ningxia and Gansu represent 24 per cent, whereby these seven provinces account for 90 per cent of China's coal reserves. However, despite the richness of resources, the contribution of coal to China's energy security is far from granted. If the primary challenge of every energy security strategy is to guarantee equilibrium between supply and demand, then in the case of coal, challenges are represented by the existing imbalances between demand and supply, as well as by the rock's environmental burden.

The transition from a planned economy to a market-oriented one generated an unprepared relationship that grew between the development of free market and the reduction of government control, rendering current coal management distant from a viable model. The imbalances produced by this relationship are particularly evident in the current system of coal pricing, incorporating state intervention and the constraints set in order to adapt to international market prices when it comes to price fixation. This translates to the difficulties that companies meet in keeping their production objectives, as coal prices and available quantities are subject to excessively strong fluctuations. Behind the scenes of this struggle to progressively adapt to a market system and rationalise coal consumption, vested interests of ministries, state-owned enterprises and local governments are significantly hindering the much-needed reforms of the sector where coal prices are the main battlefields. A reform of the ownership structure of power companies that are overly at the mercy of non-economic decision-making systems would be an important step for China's energy security.

In line with a management system that is incompatible with the rational consumption of resources, which is essential to energy security, an increasing challenge in the use of coal is the high level of pollution provoked by its production and combustion, pushing the government to review its consumption objectives in the years coming. The Chinese government is increasingly paying attention to the environmental problems caused by coal combustion, and the control over carbon emissions incites current policies to move towards a gradual decrease in its consumption. The energy development goals that the National Energy Administration promulgated in 2012 anticipated a 5 per cent decrease in coal consumption by 2015, compared to that of 2010 (Feng 2012). A reduction is also indicated in the *Air Pollution Prevention and Control Action Plan* (State Council 2013), predicting that coal consumption will pass from 72 per cent in 2012 to 65 per cent in 2017.

As we will see through the different sections of this chapter, the place of coal in the Chinese energy security strategy is fundamentally challenged by its governance system. In particular, energy security will be observed through a long-term coal demand-supply balance, analysed from a double perspective: the direct coal utilisation and that of coal-transformation in the chemical industry (Figs 2.1 and 2.2).

In Section 1, we approach the subject of the 'vertical' industrial chain conflict in coal-electricity, the most representative case to explain the primary barriers of today's coal market development. In Section 2, we look at the issue from a 'horizontal' regional perspective, analysing the predicament in coal transport, local governments protectionist policies and other non-market factors that influence the coal market. In Section 3, we explore the impact of both large-scale coal imports and the sharp reduction in coal prices on the Chinese coal market. Last but not least, Section 4 provides an analysis and comments on the Chinese government's key policies in the coal industry. We clarify the long-term potential threats in Section 5.

Adopting a vertical industrial chain perspective: the 'double-track pricing system'

From an industrial chain perspective, a major source of problems in China's energy security has for years been the so-called 'coal–electricity' nexus, particularly represented by the frequency of energy shortages throughout the last decade. The situation was to some extent eased in 2012 thanks to the fall in coal prices. The reason behind the electricity shortages was the difficulty for several coal-fired power plants to buy coal, either due to the

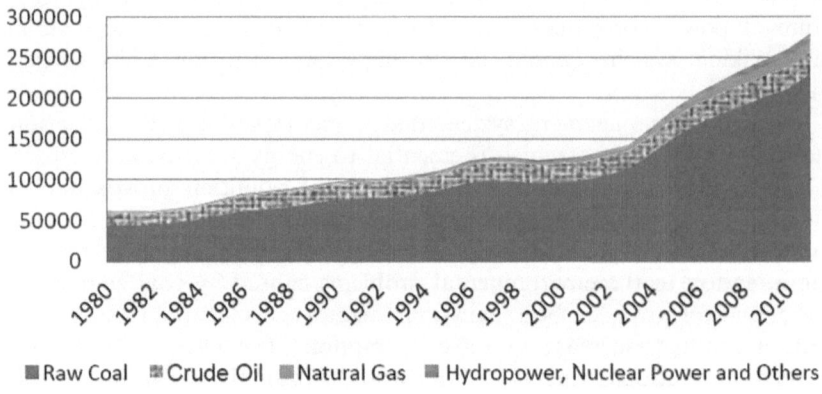

Figure 2.1 Primary energy output and breakdown
Note: Unit, 10,000 tonnes of Standard Coal Equivalent.[1]
Source: Chinese Energy Statistical Yearbook.

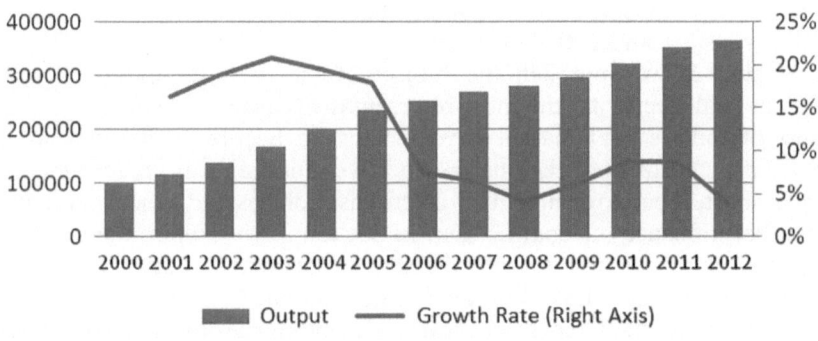

Figure 2.2 Raw coal output
Note: Unit, 10,000 tonnes.
Source: China Statistical Yearbook.

high costs or the impossibility of finding enough available resources on the market. For these reasons, power plants were obliged either to withhold their operations or to stop operating altogether, significantly impacting economic production and the everyday life of the people.

High coal costs are linked to price reform policies that have been adopted since the 1990s. In line with its intentions to introduce market reforms, the

Chinese government decided to cancel the government-fixed price of coal, together with a loosening of coal price regulations. From 1994 onwards, except for the price of coal earmarked for power generation, coal prices for other usages were determined by the market. This situation led to the creation of a double-track pricing system for coal resources. In the case of power generation, the rate of electricity was set by the government in the hope of maintaining relatively stable electricity prices. It was thus an incentive for the government to regulate the coal rate of power generation. In general, the Chinese government therefore established the possibility for power generation companies to buy coal at very low prices against the general trend of the soaring coal market price. This type of arrangement, called 'contract coal' (in Chinese *hetongmei*), allows for major coal enterprises to sign contracts with power plants according to a government guided-price, which is generally low. However, this option is quite limited,[2] and power plants ultimately also have to purchase resources at current market prices, the second 'track' of the pricing system, which in Chinese is called 'market coal' (*shichangmei*) (Fig. 2.3).

The market price of coal has been rising since 2002, entailing an important widening of the gap between this one and the government-guided 'contract coal' price. For instance, in 2011, the price of 'market coal' was 290 RMB/ton higher than that of 'contract coal' (Wei 2011). Such a difference between the two types of prices and the limited role 'contract coal' plays in supplying resources to power plants had important consequences for the power sector and the country's energy security.

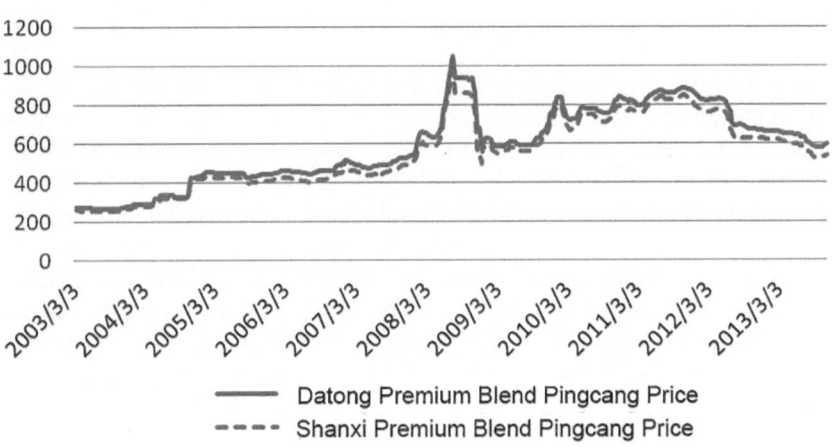

― Datong Premium Blend Pingcang Price
---- Shanxi Premium Blend Pingcang Price

Figure 2.3 Coal price trend
Note: Unit, RMB/Tonne.
Source: http://finance.ifeng.com.

Persistence in price imbalances

A first reason for electricity shortages is the difficult situation in which power companies operate due to incoherence in energy pricing systems. Since the price of electricity remains fixed by the government and is increased at a much slower pace than the coal market price, power companies have seen an important reduction in their profits over the past decade, while some have even suffered financial losses. In 2008, such price discrepancy eventually caused 42 per cent of power companies to suffer important losses (Liu 2008). For this reason, leading electricity companies started becoming more sensitive to power generation costs, largely preferring to buy coal at the low 'contract' price. On the other hand, the great price gap between the two forms of supply made coal companies more willing to sell coal at the market price, resulting in a sharp decrease in the numbers of 'contract coal' agreements and in the fulfilment of their agreements. This situation eventually led to a crisis in the supply of 'contract coal'. In order to obtain more profits, some coal companies even violated coal price policies, unilaterally deciding to increase the price of 'contract coal'. Worse yet, they also adulterated the quality of coal, while selling it as 'quality coal' in order to make more profits (Ye 2008). Finally, the great price gap also engendered important speculation regarding the 'contract coal' industrial chain.[3] For instance, in 2011, 30 per cent of 'contract coal' (about 230 million tons) was resold as 'market coal', which made intermediaries gain a profit of nearly RMB 100 for the sales of coal obtained through government fixed prices (Peng 2011).

As a result of these price discrepancies and of a general lack of control over the practices of coal companies, coal-fired power plants have to buy coal at the market price, as they are unable to procure low-price 'contract coal' as planned. If the regulation of electricity prices is maintained, then it is clear that power plants will further see their profits deteriorated in the future. This situation would oblige some companies to operate at a loss or even to shut down. Between 2004 and 2011, operating hours of thermal power facilities reduced from 5,988 in 2004 to 5294 in 2011, suggesting that some companies had indeed ceased to operate (SERC 2006–2011). Taken from an energy security and supply perspective, the imbalance created by the 'double-track coal pricing system' led to this chaotic situation in the coal-electricity industrial chain, imposing an important adjustment to electricity prices in the near future. However, challenges to energy security cannot only be ascribed to the incoherence in the pricing policy of the central government. A regional perspective looking at the role of provinces must also be taken into account.

The horizontal regional perspective: 'imbalanced endowment'

The problems encountered by the thermal power industry cannot only be attributed to diverging pricing systems, as they also derive from imbalances between supply and demand at the regional level, which could be called the 'horizontal' factors. This dilemma is not specific to the power generation industry, but it is also reflected in many other coal-consuming industries. The reasons for such 'horizontal' imbalances are multiple.

First of all, there are constraints on coal transportation that are specifically due to insufficient transport capacities. This situation in turn pushes trans-portation prices up and results in an unreasonably high price of coal. China's coal reserves are concentrated in northern and northwestern China, while the majority of coal consumers are located in the developed coastal areas of Southeast China, which means that coal has to be transported from north to south and from west to east. Coal transportation primarily relies on railways, counting for 60 per cent of China's transported coal. Similar to the presence of a double-track coal pricing mechanism, China's railway transportation is also divided into two types, one being the 'wagon within the plan' (in Chinese *jihuanei chepi*) and the other the 'wagon beyond the plan' (*jihuawai chepi*). In the first case, the carrying capacity for coal is guaranteed within the plan, while the transportation tariff is regulated by the government (and is kept quite low). In the second case, the company shall submit applications for coal transport to the railway bureau and subordinate its transportation plan to the unified dispatch of the railway departments. However, because of the limited transportation capacity, coal transportation demands cannot be guaranteed. As a consequence, and as recently as 2011, less than 40 per cent of the coal transport applications from the three main coal production areas, the 'three XIs' (the three western regions)[4] to central China, southern China and southwestern China, have been accepted (Han 2011), while rent-seeking and arbitrage phenomena became rampant in the distribution of the 'wagon beyond the plan' transport capacity.[5] As a consequence, coal transportation costs climbed even further up. An example of such an increase in transpor-tation costs can be seen in the following table, which shows data released by the 2012 annual report of Shenhua Group Corporation Limited, the largest coal producer in China (Fig. 2.4).

To worsen the situation in the coal sector, the coal price in China went down substantially in 2012, while the 'mine-mouth price' – the price of coal sold at mine sites – of 'market coal' even dropped below the 'contract coal' price. Furthermore, as can be seen in the table, the transportation cost of 'contract coal' is far less than that of 'market coal', and the difference between prices is due to the limited transportation capacity of 'market coal'. Moreover, the share of the transportation cost in coal sales prices, as shown by an approximate calculation, is such that, for 'contract coal', the railway transportation tariff takes up roughly 27.1 per cent of the off-rail coal price, while that of sea freight is about 60.7 per cent. The two figures are much higher for 'market coal', where they represent 66.8 per cent and 75.5 per cent respectively.

Secondly, another factor generating 'horizontal' imbalances is linked to the protectionist policies of local governments. These measures are taken

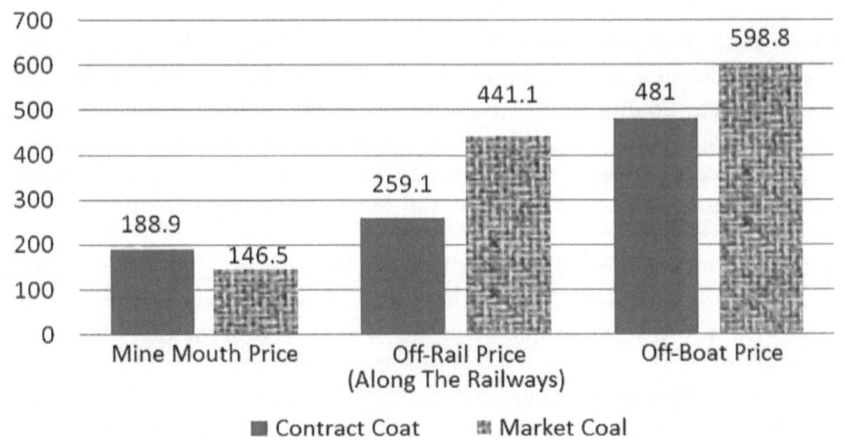

Figure 2.4 Shenhua coal sales price table
Note: Unit, RMB/Tonne.
Source: Shenhua Group Corporation Limited 2012 Annual Report.

according to the specific situation of coal supply and coal price. For instance, when coal supply is inadequate and prices goes up, local governments take measures to prevent the export of local coal to other provinces in order to guarantee local supply. An example is the policy that was adopted in 2010 in Henan province, clearly stressing that except for some key customers in other provinces, it was forbidden to sell coal outside of the province.[6] Another example comes from Guizhou province, where coal outflow was blocked by means of a 'coal out-of-province fee', requiring to pay 200 RMB/ton for coal sold to other provinces as a coal price adjustment fund.[7]

These kinds of protectionist measures are also taken when there is over-supply. This situation leads to a reduction of coal price, and local governments restrict its import from other provinces to ensure the consumption of local production. For example, in order to promote the consumption of local coal, Henan province established power generation capacity quota, encouraging companies to produce a certain level of electricity output. The government first established a coal consumption base for power generation companies; then, if the utilities purchased 10,000 tons above the established base of local coal, they would be rewarded with a 10 gigawatt basic electricity generation quota. Conversely, if their consumption were lower than 10,000 tons of coal produced within the province, a 12 gigawatt electricity generation quota would be deducted (Wang 2013).

Thirdly, in addition to governmental interferences, some other non-market factors also had a significant influence over coal demand and supply. For instance, some years ago, local governments of several coal-rich/affluent provinces took the lead to reform the coal industry, mainly focusing on the integration of coal resources, mergers and the reorganisation of coalmines. These measures led to the closure of several coalmines, as well as to the establishment of large-scale coal bases and large integrated coal enterprises. For instance in 2008, Shanxi Province started to reform the sector, which led to a drop in the number of coalmines two years later, passing from 2,598 in 2008 to 1,053 in 2010. Coalmines that had a production capacity below 300,000 tonnes per year were shut down. As a result, the number of coal companies was reduced from over 2,200 to just 120, as they were predominantly companies with yearly production capacities above 3 million tonnes (Shanxi Coal Industry Bureau). Taking Shanxi's example, other provinces proceeded with these reforms and also followed the targets established by the 12th Five-Year Plan for the Coal Industry (NDRC 2012). According to the plan, the number of coal enterprises in China should be inferior to 4,000 by 2015, while their production capacity cannot be inferior to 1 million/tons per year, clearly privileging large-scale production bases.

With the advancement of coal resource integration, the upper stream of coal market is gradually controlled by local state-owned coal enterprises, while the downstream market, namely the electricity industry, has been traditionally monopolised by central state-owned enterprises.[8] The reason why coal and electricity could not compromise with each other is mainly because, in a broad sense, both sides are state-owned enterprises (SOEs) and backed by the government, while the difference lies in that local SOEs belong to the local government and central SOEs belong to the central government. Investment returns and taxes turned in by local SOEs go to the local finance bureau and those generated by central SOEs get sent to the Chinese Ministry of Finance. Therefore, the presence of discord between the coal and electricity industries could be regarded as a conflict of interests between local and central governments. For local economic development and fiscal reasons, local governments tend to support and protect the interests of local coal enterprises, whereas the central government aims at the national allocation of coal resources to assure the coal consumption of power plants all across China, centrally owned utilities in particular.

In 2012, the situation partially changed when China's economic growth slowed down and coal imports increased, leading to an oversupply of domestic coal and a reduction in the price of coal. The constant decrease in coal prices reduced the price gap between 'contract coal' and 'market coal', sometimes lowering the price of 'market coal'. Given this situation, at the end of 2012, the State Council promulgated the *Guidance of Deepening the Reform of Electricity Generation Coal Market* (State Council 2012), cancelling the double-track pricing mechanism of coal for electricity generation. Thanks to this decision, coal and power generation enterprises are now free to sign coal purchase and sales contracts and determine coal prices through negotiation, introducing some market elements that partially ease the non-market

distortions. However, without a mature coal market, and under the circumstances that the downstream electricity price is still regulated by the government, it is beyond a doubt that abolishing 'contract coal' would lead to new
problems and risks.

Coal imports: another variable in the energy security equation

As we have seen, the incompatibility between the coal and electricity industries
is mainly attributed to systemic factors that influence coal demand and supply.
Beyond these framework factors, the significant amount of low-price foreign
coal flooding into the Chinese market accompanied by a decreasing demand in
coal triggers other changes and reversions in coal demand and supply. During
these years the sector indeed witnessed a change from high coal prices and
supply shortage to a situation characterised by low coal prices and difficulties
in selling domestic coal. The replacement of domestic resources with coal
imports unquestionably helped to improve the performances of coal-fired
utilities. However, this choice presents huge pressure and challenges to China's
coal companies, entailing a new 'energy security' issue in China's coal industry.

Since 2000, China's volume in coal imports has been increasing (except in
2008). In 2009, it surpassed that of its export for the first time; in 2011, China
became the world's largest coal importer. The volume of coal imports had
increased by 2.3 times from 2009 to 2012, and the ratio of coal imported in
national consumption volume has been rising up and even reached 8 per cent
in 2012 and close to 10 per cent in 2013 (Figs 2.5 and 2.6).

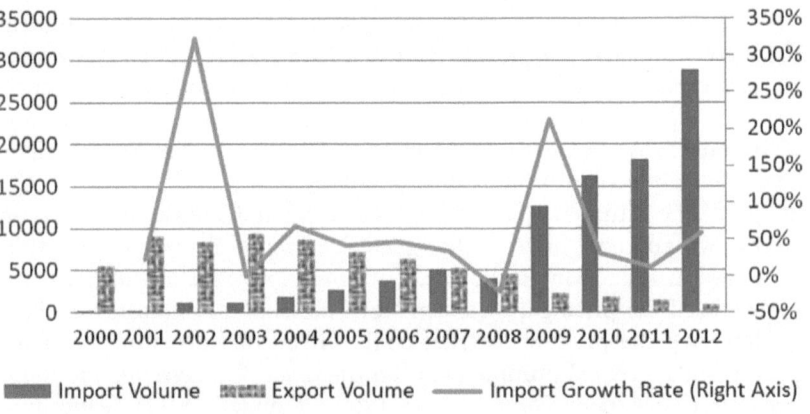

Figure 2.5 The import and export of coal
Note: Unit, 10,000 tonnes.
Source: China Statistical Yearbook.

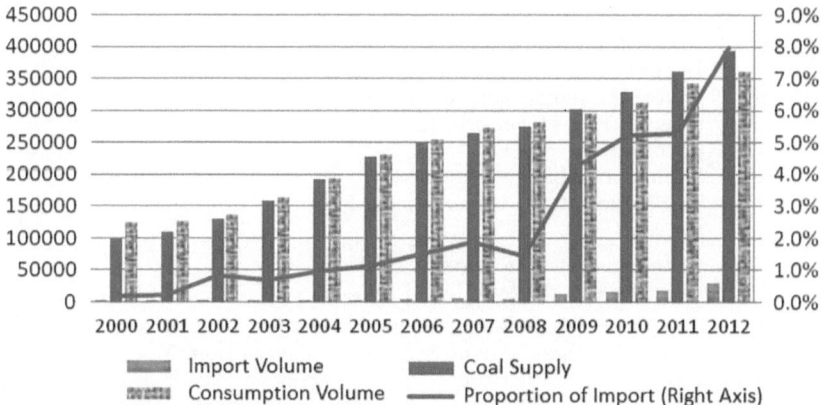

Figure 2.6 Coal equilibrium table
Note: Import ratio refers to the proportion of import volume in consumption volume.
Source: China Statistical Yearbook Statistics in 2012 are arranged by the author.[9]

In recent years, the growth of the coal imports has mainly been due to their changes and export policies. In the 1970s, in view of creating foreign exchanges reserves, China used to encourage coal exports. Lately, following the country's needs for economic development, the domestic demand for coal soared, whereas supply increased at a slower pace (Table 2.1). This situation eventually required an important change of supply strategy in 2004, which brought the government to restrict coal exports and encourage its imports. In the same year, the government released the Administrative Measures of Coal Export Quota, limiting coal export operation rights only to four companies (still valid to date). In 2006, the government decided to withdraw export tax rebate, and in the following year it imposed an export duty. Following a gradual reduction, in June 2007, the import tariff rate reached zero (except for lignite[10]), leaving the government only to levy value-added tax and port charges.

These measures of encouraging coal imports significantly increased foreign coal importance in China's domestic market, both in absolute and relative numbers. Given that imported coal had a lower price compared to domestic coal (and had the same quality), some coal-producing provinces such as Shanxi started to import coal in large quantities. Because of this price margin other provinces followed, leading to a significant reduction in domestic producers' sales and general operation difficulties for the coal industry. For instance in the first half of 2013, 45 per cent of Shanxi's coal enterprises suffered losses and over 10 per cent of them stopped production (Liang 2013). Some struggled to survive, relying on loans to pay salaries.

Table 2.1 Summary table of coal export and import policy

	Export	Import
2004–01	Export tax rebate rate of steam coal and anthracite reduced from 13 per cent to 11 per cent; Export tax rebate rate of coke reduced from 15 per cent to 5 per cent; Export tax rebate rate of coking coal reduced from 13 per cent to 5 per cent.	
2004–05	Export tax rebate for coking coal and coke was removed.	
2005–01	Export tax rebate rate of steam coal and anthracite reduced from 11 per cent to 8 per cent.	Import tariff of coking coal reduced from 3 per cent to 0 per cent.
2005–04		Import tariff of steam coal and anthracite reduced from 6 per cent to 3 per cent.
2006–09	Export tax rebate for steam coal and anthracite was removed.	
2006–11	A 5 per cent export tariff was levied on coking coal and coke.	Import tariff of steam coal and anthracite reduced from 3 per cent to 1 per cent; Import tariff of coke reduced from 5 per cent to 0 per cent.
2007–06	Export tariff of coke increased from 5 per cent to 15 per cent.	Import tariff of steam coal and anthracite reduced from 3 per cent to 0 per cent.
2008–08	Export provisional tariff of coke increased from 25 per cent to 40 per cent; Export provisional tariff of coking coal increased from 5 per cent to 10 per cent; A 10 per cent export provisional tariff was levied on other bituminous coals.	

Source: Arranged by the author based on policy announcements of the Chinese Ministry of Finance and the State Administration of Taxation.

More importantly, such a soar considerably impacted on China's coal price, and to a large extent lead to a coal price crash in 2012, when coal passed from 805 RMB/ton to 635 RMB/ton.[11] Though other reasons may also have had an influence on coal prices, such as the slowdown in China's economic growth, the restrain on high energy-consuming industries, the slower speed of growth of national electricity consumption, the weakening coal demand and the domestic coal output increase, imported low-price coal greatly effected the domestic market. Conversely, as coal prices dropped, electricity enterprises started witnessing a drastic growth in profits.

From the perspective of electricity enterprises, coal imports generally ease coal supply pressure and relieve the tension of electricity supply in the eastern coastal areas. However, for coal enterprises, the import of the rock presents a significant impact on the nationwide industry, including that large numbers of domestic coal enterprises were in deficit and on the verge of bankruptcy. These enterprises hence demand a restriction of the import from the state level and thus generate much uncertainty regarding the domestic coal supply.

Ensuring energy security while solving coal-electricity contradictions

In order to solve these pressing problems in the coal and power generation sectors and to maintain an adequate balance between supply and demand for energy security concerns, the government has over the last decade promoted a series of new policies. In particular, it established a linkage between coal and electricity prices and new 'vertical integration policies' for both sectors, beyond coal reserves mechanisms.

The price coordination policy: coal-electricity price linkage

At the end of 2004, the government announced the vertical pricing system reform policy in order to address the coal–electricity conflict, linking the price of electricity and the price of coal. This policy established that when coal prices rose to a certain degree, then electricity prices would be adjusted accordingly. After its promulgation, the first (May 2005) was implemented as planned, whereas the second adjustment was put off for nearly a year, to July 2006. The policy was finally suspended because of fears that the rise in electricity prices might lead to inflation and trigger public discontent. From that moment on, electricity prices were directly regulated by the Chinese National Development and Reform Commission (NDRC).

In December 2012, the State Council issued the Guidance of Deepening the Reform of Electricity Generation Coal Market by State Council General Office (State Council 2012), putting forward a new coal–electricity price linkage in response to the fluctuations in the price of coal and stipulating that the price of electricity shall be adjusted based on a one-year cycle. The outcome of this policy is still far from known.

Firstly, the coal–electricity linkage policy cannot guarantee economically reasonable prices of coal supply. In essence, the linkage is a response to the surge in coal prices. The mechanism ensures that power companies purchase coal. However, according to the pricing formula, the rise in coal prices will be transmitted to electricity consumers, resulting in the relative decreasing sensitivity between coal prices and demand. In other words, the coal–electricity price linkage indirectly forces electricity consumers who have little say in pricing issues to shoulder the coal price increase rather than to ensure a reasonable price in coal supply, which is against our definition of coal energy security.

Secondly, the coal–electricity price linkage is not conducive to solving the coal–electricity conflict. In whatever circumstances, the adjustment of electricity prices is always far behind the fluctuations in coal prices in terms of timing and scope. Moreover, the vertical pricing system reform as such fails to touch the aforementioned problems such as the electricity coal price dual-track system and the cross-region coal circulation barriers.

Coal–electricity vertical integration and stronger monopolies

The long-standing coal–electricity vertical integration,[12] whose main function is to guarantee coal supply to local power plants or to help local coal companies in case of oversupply, was clearly put forward in the *Several Opinions for Acceleration of Coal and Mineral Enterprises' Merger and Restructuring* (NDRC 2010) promulgated by the NDRC in 2010, and has since become a long-term policy.

With the stringent government control over electricity prices and relatively flexible coal prices, the coal–electricity joint operation enables larger coal sales when the coal price is high, and more electricity supply if the coal price is low. As such, the operation risk of small enterprises is minimised. Besides, the coal–electricity joint operation mechanism and the establishment of pithead power plants will help reduce coal transformation fees, which is a significant part of the electricity generation cost.[13]

Nevertheless, it should be noted that the coal–electricity joint operation in China is not a voluntary market choice made by electricity and coal enterprises, but it is rather advocated by government administrative intervention. A good example is the role of the Shanxi government in the Datong Coal Mine Group's acquisition of Zhangze Electric Power.[14] Administrative intervention of this kind can cause many problems.

To begin with, no-one can expect to solve the electricity–coal supply shortage through the coal–electricity joint operation. When the coal price goes up, enterprises are prone to selling coal and generating less electricity because it is more profitable. In 2011, thermal power plants of China's five major electricity generation groups witnessed a total loss of RMB 31.2 billion. In 2011, China's total coal yield was 226 million tons, with the average coal self-sufficiency rate of around 25 per cent. According to the State Electricity Regulatory Commission report, most of the coal had been sold directly.

In addition, the coal–electricity joint operation reinforces monopoly and unfair competition. Even if coal purchased by the give major power giants is internally sold for electricity generation, it raises the possibility of price discrimination and trade barriers, such as prioritising purchases and sales of their own coal. The strengthened monopoly and unfair competition will jeopardise the market, particularly when the coal price slumped against surging electricity prices. What is more, the 'coal-electricity joint operation' enterprises have an unfair advantage over other coal enterprises. Furthermore, on the occasion of rising coal prices and stagnating electricity rates, 'coal–electricity joint operation' enterprises can easily knock down electricity plants. All in all, the coal–electricity joint operation will be disadvantageous to both industries and could also harm the interests of consumers.

Coal reserve mechanism

In 2011, an important move to guarantee a balance in coal supply–demand was the release of the Interim Measures on the National Coal Emergency Reserves (NDRC 2011a) by the NDRC. These measures mark the build-up of China's national coal reserve mechanism, helping the government to restrain coal price fluctuations via coal reserves and to maintain market stability.

The main function of China's coal emergency reserves is to provide priority to the State to purchase coal from large-scale companies in the case of an emergency. Under this mechanism, large-scale companies function as storage spots, implying that they are allowed to buy and sell coal with the obligation of prioritising the reserve target stipulated by the government, while the latter supports these companies by means of investments and interest subsidies.[15] However, this current coal reserve policy raises a series of doubts.

According to the above-cited document, China's coal reserve mechanism is still on trial, so the reserve remains quite small, and its target for 2013 is 6.7 million tons. In 2012, the target was 5 million tons, which only covered 0.17 per cent of the total consumption that year. If kept in such a low quantity, the reserve can hardly stabilise the commodity price and ensure coal supply.

In addition, because stored coal can suffer wind abrasion and deterioration, coal reserves may entail serious environmental pollution, while needing a supply of land and storage expenditure. Therefore, further investigation and verification are needed to determine an optimum reserve target. Doubts emerge concerning the long-term development of the reserve mechanism. Coal reserve bases are currently foreseen as depots only, thus reducing the profits of the coal companies in question, a situation that is hard to sustain unless stocks are financed by significant government subsidies.

From this graph, it is clear that problems in China's coal industry are mainly related to non-market system factors. Unfortunately, government steps towards market liberalisation have been quite few, given that administrative measures are always preferred when it comes to solving such problems. However, while on the one hand the central government wishes to solve existing conflicts between coal companies and power generation companies, as well as the coal protectionism of provinces, on the other hand these further 'distortions' bring new challenges and leave fundamental issues untouched. In order to guarantee a better balance between supply and demand, the government surely cannot exclusively rely on the market's role, even though a regulation role is always necessary. However, interventions in the market should not replace the market itself, but rather play a perfecting and complementary role.

Alternative use of coal for an energy security strategy

In China, the coal chemical industry is believed to efficiently process and convert coal, as well as utilise it to its utmost potential. Coal liquefaction and coal gasification techniques are also regarded as alternative strategies to oil and natural gas, possibly helping China to reduce its reliance on foreign oil and gas supply. Yet the current development of China's coal chemical industry is not satisfactory for a series of reasons: an overcapacity of traditional coal chemical products, unclear prospects of modern coal chemical industry and huge challenges to the already deteriorating environment all cast important doubts over the development of coal chemistry.

Overproduction in the traditional coal chemistry industry

Frantic development and overcapacity are currently major problems in China's traditional coal chemical industry, a situation that resulted in the 'race' of almost all coal-producing provinces and certain coal-importing provinces in the aim of developing traditional conversion of coal facilities, less demanding in terms of technology and investments than modern coal conversion. In 2008, 20 or so provinces and autonomous regions planned to 'forge' the coal chemical industry as their pillar industry (Pan 2008), resulting in an important soar in the production capacity. In 2010, the production capacities of synthetic ammonia, methanol and calcium carbide represented 35 per cent, 50 per cent and 97 per cent, respectively, of the world's total (NDRC 2012).

To contain this blind planning and development of coal chemical projects, the NDRC released numerous documents,[16] demanding stricter examination and approval, as well as an upgrade of access standards.

Government regulations have until now failed to slow this rapid development of the coal chemical industry down. The main reasons are to be found in the low technical barriers for the traditional coal chemical industry, the relatively mature development of the sector and the small investment scale. Most importantly, the principal factors were surely the desire of local governments for GDP growth coupled with administration performances indicators and local protectionism. Provinces rich in coal resources in particular wish to increase their profits by extending the industrial chain. Many of them establish rigid requirements regarding the on-site conversion rates of coal[17] in order to block the outflow of local coal resources and force coal companies to build up coal chemical projects to obtain coal resources.

To further complicate the issue, several coal-importing regions where thermal power coal mainly relies on the supply from other provinces also started to become actively involved in the coal chemical industry, threatening the local coal demand–supply equilibrium even more. According to incomplete statistics, the quantity of coal consumed by coal chemical projects under construction and upon approval throughout the country exceeded 100 million tons at the beginning of 2011 (NDRC 2011b). If other projects currently being planned are taken into account, then several more million would be added up. It is then foreseeable that the development of the coal chemical industry worsens coal supply issues, whereas overcapacity and the limited raw material supply will intensify the competition among coal chemical enterprises, where some companies might be forced to shut down.

A still young and modern coal chemical industry

China's modern coal chemical industry is still at its early stage of development since investing in such a technology-, capital- and resource-intensive industry is very risky. In this regard, the Chinese government has initially adopted a cautious stance, first preferring to launch a pilot project and promote business development once the first trials have been evaluated. However, since the end of 2012, the Chinese government has gradually loosened its regulation and decided to approve many projects in order to respond to the air pollution crises.

A slow development of substitute natural gas[18]

In recent years, the increasing domestic natural gas demand and rising gas price has triggered enthusiasm in coal-to-gas projects. However, this industry still lacks definite government industrial policies and is still emerging. After a period in which the NDRC had been limiting the approval of coal-to-natural gas projects and their supporting projects from local governments,[19] the worsening of air conditioning in China in 2012 brought the government to change its stance. The NDRC gradually started to loosen the examination and approval of projects, and in the first seven months of 2013, 15 projects were approved with a gross capacity of 52.284 billion cubic meters/year (Yu 2013) along with hundreds of billions of investments. It is expected that in 2015 the coal-derived synthetic natural gas supply will amount to 15 to 18 billion cubic meters, accounting for 8.5 per cent to 10 per cent of domestic natural gas yield (NDRC 2012).

However, the massive development of coal-to-gas projects faces similar problems as the development of the gas industry, as it is severely hindered by the construction and accessibility of the natural gas transportation network, which the big national oil and gas production companies are currently monopolising. There are two fundamental solutions to the transportation problem. A first one would be the construction of new pipeline networks, a possible choice for rich state-owned companies such as Datang Corporation or the China National Offshore Oil Corporation (CNOOC). However, the precondition for self-built pipelines is to have long-term gas purchase, which means that the profits of the target market could cover the investment costs. Moreover, since the production capacity of certain single projects is quite limited, the pipelines are not distant from the target market; otherwise the investment and operation cost would be overly high. A second option would be the construction of transportation pipelines in the main production areas of coal-derived synthetic natural gas, including Inner Mongolia Province and Xinjiang Province. This option is currently planned by Petrochina and SINOPEC. In this case, producing companies may choose to sell synthetic natural gas to these two big groups, though with significant disadvantages in the face of their monopoly. Their investments are thus bound to being negatively affected.[20]

Another factor that impedes the development of coal-derived synthetic natural gas is the lack of reform in the natural gas industry. The development of China's coal-derived synthetic natural gas industry is currently subject to the changes of the whole gas industry. Implying its possible marketisation, a reform of prices in the natural gas industry could be at the basis of a reduction of the overall gas price, whereas the commercialisation of non-conventional natural gas such as shale gas would also have an impact on the development of the coal-derived synthetic natural gas industry. In sum, the obstacles for the development of the coal-to-gas industry are still very important.

The development of coal liquefaction: unclear advantages[21]

With an increasing domestic demand for refined oil, as well as a rise in international crude oil prices, many coal companies have in recent years started to turn their attention towards coal liquefaction projects. However, regards the coal-to-gas projects, the NDRC decided in 2008 to withdraw examination and approval rights, which impeded the achievement of the target capacity of 30 million tons/year in 2010 as foreseen by the initial plan of the *Medium and Long-Term Development Plan of Coal Chemical Industry*. Except for two pilot programmes developed by the Shenhua Group,[22] the examination and approval of other coal liquefaction projects were suspended without exception.[23] Lately, with the coal price reduction of 2012, the examination and approval policy were unfastened and a dozen projects were approved.

However, there are enormous controversies surrounding the large-scale proliferation of coal liquefaction techniques, in particular concerning its excessive consumption of coal, water and energy (in particular hydrogen) (Chen 2013) and the related added carbon emissions. Moreover, its energy conversion efficiency is the lowest among all coal chemical products: major coal liquefaction products are liquid fuels like gasoline and diesel, where the efficiency of energy conversion is quite low when used as vehicle fuels (Xu 2012). Coal liquefaction is therefore not entirely suitable for large-scale development. Besides, the quality of the derived diesel is not satisfactory and can only be used after further mixture and processing. However, Tang Hongqing[24] claims that simple comparisons of advantages and disadvantages of techniques do not make sense, since different coal liquefaction techniques produce various kind of products that are suitable for different usages. In addition, since the techniques used for coal liquefaction and other coal chemicals are complementary, the combination of these different techniques is said to greatly improve the overall efficiency (Zhang 2013).

A worrying development of the modern coal chemical industry

In addition to the particular problems encountered by coal-to-gas and coal liquefaction techniques, there are various other concerns over the development of China's modern coal chemical industry.

First and foremost, the coal demand–supply balance may be threatened. The modern coal chemical industry is coal-intensive, while projects under construction are all large scale, with production capacities that represent several million tons. The development of this sector would thus imply a high pressure over coal supply, further adding existing concerns and entering into competition with other types of coal-consuming industries.[25]

Secondly, as the price of coal is currently quite low, the coal chemical industry development is surely encouraged. However, given that the process of project approval is considerably lengthy and can last for at least two years, there are significant risks of coal price fluctuation. Compared to traditional coal transformation, modern coal chemical projects imply larger construction scales and greater investments, as well as supporting facilities and utilities such as railway lines, water sources and electricity (specialised for the use of coal mines). In the case of overcapacity, it will be impossible to reach the expected profits, entailing profound damages not only for the company that invested, but also for the local economy, and society as well as the banking and insurance systems, the consequences of these uncertainties can therefore be disastrous.

Furthermore, and as was already pointed out, the large-scale development of the modern coal chemical industry presents serious constraints on water resources, especially in arid coal production areas. The settlement of water-consuming coal chemical projects in main coal-producing provinces of Western China would worsen their hydrological situation, pushing them to divert water from the Yellow River or its tributaries (while the annual water quotas allocated to these provinces by the Yellow River Conservancy Commission are somewhat limited). This situation would further exacerbate cases of 'stolen' water from the Yellow River or of overexploitation of ground-water, all the while threatening the drinking water supply of residents (Guo 2012).

Besides high water consumption, there is also the serious problem of large amounts of wastewater discharge. Companies could opt for recycling water pollution treatment, but this is difficult to implement and not economically sound. Only one coal chemical wastewater 'zero-discharge' project has been approved by far, the Shenhua Coal Liquefaction Project. However, Greenpeace recently showed how this project not only discharged severely polluted water, but also illegally pumped large amounts of groundwater, resulting in irreparable damages to the local environment.

Finally, the considerable amount of carbon emissions on behalf of the coal chemical industry also raises concerns. According to calculations,[26] the production of one ton of fuel by indirect coal liquefaction would release 6.1 tons of CO_2, whereas it would be 5.8 tons of CO_2 if the direct coal liquefaction technique were used, and the production of 1,000 cubic metres of gas would release 4.8 tons of CO_2. The operation of large-scale modern coal chemical projects would thus add further pressure to China's already very significant carbon emission.

Conclusion

We contend that in the case of coal, the most crucial threat to energy security is the imperfect market and administrative interference. In recent years, the outstanding problem of coal security has been the presence of a structural conflict between coal and downstream enterprises (primarily coal-fired power plants) and railway transport departments. There are important conflicts of interest for the different stakeholders in the coal chain, as most coal enterprises are locally owned state enterprises, while coal-fired power plants are mostly centrally owned. Meanwhile, railway operation is frequently intervened by administrative power. In face of problems in the coal industry, both central and local governments issued too many administrative policies to intervene in the market, resulting in the structural conflict over coal supply and triggering the phenomena of local protectionism, although the large amounts of coal imported from foreign countries and the decreasing coal prices have to a certain extent eased the coal supply tension and relieved the coal–electricity conflict. Notwithstanding, the essential issues remain unsolved and thus pose a great threat to coal security.

Coal resources are owned by the local government, which has the right to contract, sell or grant coal resources to enterprises. After the government-led integration of coal resources, represented by Shanxi Province, coal resources found in Shanxi are almost manipulated by the five state-owned coal enterprises affiliated to the Shanxi provincial government. However, due to government support and market predominance, most state-owned companies are faced with the problem of high costs, overstaffing and low efficiency. This situation does not ensure an efficient utilisation of coal resources.

The marketisation reform of China's coal industry only liberates the pricing mechanism without changing the original property relations or property management mode. The ultimate solution to these problems lies in the marketisation reform of the coal industry, the establishment of an ownership system in line with market rules, and a reform in state-owned enterprises. However, these aspects are exactly where most reform obstacles come from. Reforms of the sector will surely harm the vested interest within state-owned enterprises and bring about political risks. Since the Chinese government is always under great pressure to maintain stability, few expectations can be made over governmental capacities or willingness to take such risks and thus to promote reforms.

As far as the prospects of the industry are concerned, and in particular in the coal conversion sector, reforms are also imperative in this aspect. The coal chemical industry needs to address its several inefficiencies, such as the excessive production and presence of too many small players, which is the reason behind overcapacity. As for China's modern coal chemical industry, the great development triggered by the governmental examination and approval system is very risky for the industry and may entail important threats to the coal demand-supply balance, not to mention the environmental problems related to its development. If coal is going to play an important role in China's energy security, then the challenges ahead for a rational utilisation remain numerous.

Notes

1 Computation method: Calorific Value Calculation.
2 The supply quantity of 'contract coal' is a compromise between interests of coal enterprises and power companies. Too much 'contract coal' will threaten the profits of coal enterprises.
3 Intermediaries purchase coal from coal mines or power plants (some power plants sell spare contract coal to gain profits, while others do so to make up for the loss caused by electricity generation) with a price higher than that of contract coal yet lower than the market price. They would then resell the coal to coal enterprise consumers such as power plants and chemical factories at a higher price (though still lower than the market price).
4 'Three XIs' (three western regions) refers to Shanxi province, Shaanxi province and the Western Inner Mongolia Autonomous Region, where coal reserves make up 60 per cent of the national total. They are the main production areas of coal and the coal transported from these regions to other provinces accounts for 90 per cent of China's inter-province coal transportation volume.
5 For example, during an interview in April 2011, a coal middleman mentioned that the coal price negotiated with the coal enterprise was 360 RMB/tonne. To transport them to the destination, one had to pay an extra cost of as much as 200 RMB/tonne to the railway transportation department (*dianzhuangfei* in Chinese) for railway 'wagon beyond the plan' (21st Century Economic Herald, May 2011).
6 In December of 2010 at the Coal Demand and Supply Convergence Meeting, the Henan provincial government clearly stated that except for certain key customers in Hunan and certain other places, Henan's coal could not be sold to other provinces.
7 *Administrative Measures on Levy and Usage of Coal Price Adjustment Fund* in Guizhou Province, October 2011.
8 Out of China's gross installed capacity in 2011, 49 per cent belong to the five centrally owned power generation groups.
9 The author arranged the statistics in 2012 according to the coal output, import and export volume, as well as the consumption volume revealed by the National Bureau of Statistics, without considering inventory margins.
10 Government restricted the import of low-heat and high-sulphur content lignite for environmental concerns.

11 According to the China National Coal Association, by the end of December 2012, the price of 5,500 Kcal/kg NAR coal in Port Qinhuangdao was 630–640 RMB/tonne, down 170 RMB/tonne compared to that of early 2012.

12 Coal–Electricity Integration refers to the vertically integrated operation of coal and thermal power plants along the industrial chain. To be more specific, enterprises can sell coal and electricity at the same time: either coal enterprises participate in electricity generation and sales, or power companies sell coal, or enterprises directly invest in coal mines and generate and sell electricity as well.

13 This will add to electricity transmission fees, which is generally speaking higher than coal transportation fees.

14 The acquisition was significant in that the China Power Investment Corporation and the Shanxi International Electricity Group transferred their 18,500 and 11,413 shares of equities of the Shanxi Zhangze Electric Power to the state-owned Assets Supervision and Administration Commission of Shanxi Provincial Government gratis, which later gave a carte blanche to the Datong Coal Mine Group. Within a year it was merged into the Datong Coal Mine Group by means of capital increase and the shareholding of the Datong Coal Mine in the Zhangze Electricity Power mounted to as much as 43.45 per cent. According to an expert in the matter, without the full support and meditation of the Shanxi Provincial government, the Datong Coal Mine Group could not have gone this far (Li Yangdan, August, 2012).

15 For example, the Ministry of Finance grants interest subsidies to the funds used for national coal contingency reserves and provides fixed subsidies to the management cost and rental. In addition, as the reserve period ends, enterprises can apply for subsidies by submitting reports, and the subsidy will be refunded by the Ministry of Finance upon approval (see National Development and Reform Commission, 2011a).

16 Notice on Strengthening Management of Coal Chemical Project Construction and Promoting Healthy Industry Development (2006), Restructuring and Rejuvenation Plan of the Petrochemical Industry (2009), Several Opinions on Curb Overcapacity in Certain Industries and Redundant Projects and Promoting Healthy Industry Development (2009), Notice on Standardisation and Orderly Development of Coal-Chemistry (2011).

17 For example the *Administrative Regulations on Coal Resource Compensated Configuration, Exploration, Development and Conversion of Xinjiang Province* (Interim), issued in October 2011, stipulated that the on-site conversion rate of coal resource development and utilisation for coal electricity and coal chemical projects shall reach 60 per cent or above. The Coal Merger and Restructure Work Plan of Inner Mongolia (March 2011) specified that newly built coal production projects must simultaneously launch conversion projects and that the on-site conversion rate must be 50 per cent or above.

18 'Coal-to-Gas' refers to the technology that is used to produce Substitute Natural Gas (SNG) by processing synthetic gas from coal gasification and methane.

19 The National Development and Reform Commission, *Notice on Relevant Issues Concerning Standardisation of Coal-to-Gas Industry Development*, June 2010.

20 We can see the example of the Keqi coal-to-gas project of the China Datang Corporation. This project is the first national pilot program jointly invested by the China Datang Corporation and the Beijing Gas Group, with the gross capacity of 4 billion cubic meters. The Phase I project produced coal-derived gas in July 2012. To supply gas to Beijing, Datang alone constructed a pipeline from Inner Mongolia Province to the Miyun County of Beijing. However, the pipeline from Miyun County to the centre of Beijing is monopolised by the China National Petroleum Corporation (CNPC), so Datang therefore had to first sell the gas to CNPC, but both sides failed to reach a consensus over the on-grid price. The

inauguration of Datang's gas pipeline was put off time and time again. In 2013, when Beijing was in desperate shortage of natural gas, the coal-derived gas eventually arrived in thanks to the multilateral mediation led by the government.

21 Coal Liquefaction is a general term referring to a family of processes for producing liquid fuels and petrochemicals from coal. There are two technical routes: direct coal liquefaction and indirect coal liquefaction.

22 The Shenhua Group's two pilot programmes are the Inner Mongolia Direct Coal Liquefaction Project, which was put into operation in 2008, and the cooperation program with South Africa-based petrochemical firm Sasol, which has since been suspended.

23 The National Development and Reform Commission, *Notice on Relevant Issues Concerning Strengthening Coal-to-Oil Project Management*, August 2008.

24 Tang Hongqing, senior engineer of Synfuels China Technology Co., Ltd.

25 Ningdong Base in Ningxia Province is a good example, where there is a 4 million tonnes indirect coal liquefaction project and a 3 million tonnes coal-to-olefins project currently under construction. In the next decade, the base expects to realise the annual coal deep processing capacity target of over 100 million tonnes. However, Ningxia's coal output in 2012 was only 85.98 million tonnes. Even if the coal yield increases, the coal chemical industry development in Ningxia, the province with the sixth largest coal reserves across China, will face great difficulties as well.

26 Statistics inferred from the Coal Deep-Processing Demonstration Projects Plan during the 12th Five-Year Plan period.

Bibliography

Chen, J. (2013) 'Coal Liquefaction Shall Enhance Technical Basis', *China Chemical Industry News*, June. Available at: www.ccin.com.cn/ccin/news/2013/06/14/265816.shtml (accessed 20 March 2014).

Chinese Ministry of Land and Resources (2014) *China Gazette of Land and Resources 2013*. Available at: www.mlr.gov.cn/zwgk/tjxx/201404/t20140422_1313358.htm (accessed 3 November 2014).

Cui, C. (2011) 'China's Coal Resources and Their Distribution Characteristics.' *Sci-Tech Information Development & Economy* 21: 181–182.

Feng, Y. (2012) 'China's Coal Consumption in 2015 Will Decrease by 5% as Opposed to 2010', *China National Radio*, November. Available at: www.chinadaily.com.cn/hqgj/jryw/2012-11-28/content_7626351.html (accessed 20 March 2014).

Greenpeace (2013) Thirsty Coal: An Investigation Into the Over-Extraction of Groundwater and Illegal Discharge of Waste Water by Shenhua's Coal-to-Liquid Demonstration Project in Ordos, Inner Mongolia, July. Available at: www.greenpeace.org/china/zh/news/releases/climate-energy/2012/08/coal-west/ (accessed 15 April 2014).

Guo, R. (2012) 'High Water Consumption Becomes the Greatest Potential Concern About the Coal Chemical Industry', *Economic Information Daily*, September. Available at: http://jjckb.xinhuanet.com/2012-09/18/content_401985.htm (accessed 20 March 2014).

Han, Z. (2011) 'Coal Railway Transport Capacity Will Not Relieve Before 2014', *Huachuang Securities*, August. Available at: www.yicai.com/news/2011/08/1041406.html (accessed 20 March 2014).

Liang, X. (2013) 'Coal Price Down to Cost Bottom Line, Half of Shanxi's Coal Enterprises Losing Money', *Xinhuanet*, July. Available at: http://news.xinhuanet.com/energy/2013-07/08/c_124976076.htm (accessed 20 March 2014).

Liu, X. (2008) 'Involuntary Losses of Power Companies.' *Sina Finance.* June. Available at: http://finance.sina.com.cn/chanjing/b/20080616/19004987328.shtml (accessed 6 September 2014).

Li, Y. (2012) 'Zhangze Electric Power Case Underlines Shanxi's Coal-Electricity Joint Operation Predicament', *China Securities,* August. Available at: http://news.xinhua net.com/fortune/2012-08/03/c_123520592.htm (accessed 20 March 2014).

National Bureau of Statistics, Energy Statistics Division. *Chinese Energy Statistical Yearbook 2002–2014.* China Statistics Press.

National Development and Reform Commission (2004) *Administrative Measures of Coal Export Quota.* Available at: www.gov.cn/gongbao/content/2004/content_62969. htm (accessed 20 March 2014).

National Development and Reform Commission (2010) *Several Opinions for Acceleration of Coal and Mineral Enterprises' Merger and Restructuring.* Available at: www.gov.cn/zwgk/2010-10/21/content_1727160.htm (accessed 20 March 2014).

National Development and Reform Commission (2011a) *Interim Measures on the National Coal Emergency Reserves.* Available at: www.mof.gov.cn/zhengwuxinxi/ zhengcefabu/201106/t20110602_556274.htm (accessed 20 March 2014).

National Development and Reform Commission (2012) *The 12th Five-Year Development Plan for Petrochemical and Chemical Industries.* Available at: www.gov.cn/ zwgk/2012-03/22/content_2097451.htm (accessed 20 March 2014).

National Development and Reform Commission (2011b) *Notice on Standardisation and Orderly Development of Coal-Chemistry.* Available at: http://bgt.ndrc.gov.cn/ zcfb/201104/t20110412_498601.html (accessed 20 March 2014).

Pan, L. (2008) 'Some Thoughts on China's Coal-Based Energy Chemicals Development.' *China Coal* 5: 64–68.

Peng, F. (2011) 'Coal Dealers Uncovers Truth behind Electricity Shortage', *National Business Daily,* May. Available at: http://news.ifeng.com/society/5/detail_2011_05/24/ 6591886_0.shtml (accessed 20 March 2014).

State Council (2012) *Guidance of Deepening the Reform of Electricity Generation Coal Market by State Council General Office.* Available at: www.nea.gov.cn/2012-12/26/c_ 132064264.htm (accessed 20 March 2014).

State Council (2013) *Air Pollution Prevention and Control Action Plan.* Available at: www.gov.cn/zwgk/2013-09/12/content_2486773.htm (accessed 3 November 2014).

Shanxi Coal Industry Bureau (2010) Report on Shanxi Province's Coal Enterprises Mergers and Reorganisation Work, February. Available at: www.docin.com/p -710587706.html (accessed 20 March 2014).

State Electricity Regulatory Commission, the Annual Report on Electricity Regulation 2006–2011.

Wang, L. (2013) 'Local Governments Bailing Out the Market: Promulgating Coal-Electricity Mutual Protection Policies.' *China Business Journal* (June). Available at: www.cb.com.cn/economy/2013_0608/1000724.html (accessed 6 September 2014).

Wei, S. (2011) 'Coal-Electricity Negotiation: Focusing on the Stabilisation of Coal Price', *Securities Times of China,* November. Available at: http://epaper.stcn. com/paper/zqsb/html/2011-11/17/content_321355.htm (accessed 20 March 2014).

Xu, W. (2012) 'Coal Liquefaction Shall Not Be Blindly Expanded.' *China Chemical Industry News,* December. Available at: www.ccin.com.cn/ccin/news/2012/12/03/ 247627.shtml (accessed 20 March 2014).

Ye, Y. (2008) NDRC Vows to Strengthen Tentative Electricity-Coal Price Intervention, July. Available at: http://finance.sina.com.cn/chanjing/b/20080725/03205129801. shtml (accessed 6 September 2014).

Yu, C. (2013) 'The Largest Domestic Coal-to-Gas Project Approved: SINOPEC Speeding Up Raw Material Restructuring.' *National Business Daily.* Available at: http://big5.ce.cn/gate/big5/finance.ce.cn/rolling/201308/12/t20130812_1180857.shtml (accessed 20 March 2014).

Zhang, X. (2013) 'Coal Liquefaction: Energy Alternatives, Different Routes to the Same Destination', *China Petroleum and Chemical Industry*, October. Available at: www.ccin.com.cn/ccin/6072/6073/index.shtml (accessed 6 September 2014).

3 Gas in China's energy security strategy

Threat of a new form of dependency?

Giulia C. Romano, Yin Na and Zhang Xuanxuan

Introduction: in search of cleaner energy sources[1]

In an effort to reduce its pollution emissions and respond to international calls regarding the global effort to reduce CO_2, China is progressively looking for clean energy sources so as to reduce its heavy reliance on coal. In the coming years, given that the Central government's objective for the 12th Five-Year Plan (2011–2015) is to increase the share of non-fossil fuel in the total energy mix by up to 11.4 per cent, special attention will most likely be paid to developing the renewable energy sector. However, the rapid pace of urbanisation, together with a growing demand for energy, mean that the country will have to pay closer attention to other energy sources, in particular natural gas, which is expected to play a greater role, not only in supplying household energy consumption, but also in the transportation and power generation sectors.

The potential for rapid growth in the share of natural gas in the energy mix are certainly supported by the current macroeconomic plans. By the end of the 12th Five-Year Plan, natural gas should reach 8.3 per cent of primary energy demand, and 10 per cent by 2020, while natural gas production nationally should reach 188 billion cubic metres. However, within the same period, the demand for gas is expected to grow by 230–260 billion cubic metres. The development of city gas, combined with the current rate of urbanisation and the expansion of industrial uses (including the petrochemical industry, and to a lesser degree the power sector), as well as transportation, are all driving forces behind this rapid increase. Future challenges generated by this increasing gap in supply–demand naturally call into question China's current gas strategy.

The future 'burst' in the demand for natural gas will imply an increase in natural gas imports – obviously leading to questions concerning energy dependency – but also important needs to develop alternative natural gas sources (and synthetic gas[2]) within its own national production. China has started to increase the importation of natural gas via the pipeline connection with neighbouring countries and to build liquefied natural gas (LNG) terminals. The country has also established operations to expand its domestic production, not only through conventional sources, but also by using the high reserves of unconventional gas lying beneath the surface in China. Finally, with a concentration of production in the west of the country and the bulk of consumption in the east, China plans to expand its current gas transportation network and storage capacities.

Yet these measures remain insufficient to respond to the ambitious targets for developing gas. In order to meet this growing demand, a combination of various actions on the supply side will be required, which in turn will involve a wide variety of policies and reforms on energy prices, infrastructure and production capacities, as well as developing favourable conditions for the unconventional sectors.

With regard to national infrastructure and production, insufficiency in the transportation network obliges the country to select more costly and delicate options, such as LNG terminals, which require significant investments for their maintenance. This inadequacy in infrastructure also poses important questions over the possible development of unconventional gas, where exploitation is already hindered by environmental problems and institutional 'voids'. Moreover, increasing gas production through unconventional resources will primarily lead to increasing the rate of energy price reform. Prices set by the government are not yet on a par with international market prices, which in turn lowers the chance of companies being interested in investing in unconventional sources. The price differential also creates problems for the future supply of natural gas, and national companies that are already operating without profits become obliged to sell gas at prices below import prices. The final effect is such that the use of natural gas is discouraged in order to replace other more polluting fossil fuels.

Hence there is still a long way to go before gas can play an important role in China's energy mix and at the same time respond to its energy security and environmental concerns. The elements that hinder the development of national resources are playing against energy security concerns, leading to China being forced to increase its dependence on foreign sources. We argue that in order for gas to play an important role in China's energy security strategy, it is of paramount importance that its management model is subject to significant restructuring.

Gas in China's energy mix: production, consumption and policies

Gas is a relatively abundant resource in China. The first discovery of gas reserves on the Chinese territory was in 1998, when Chinese authorities surveyed a total of 662 oil and gas fields. The total registered resources currently amount to 60,000 billions per cubic metre, with 35,000 billion being directly exploitable. By the end of 2011, the available non-conventional resources in China amounted to 50 tonnes per cubic metre, which included 36 tonnes per cubic metre of shale gas, 9 tonnes per cubic metre of coalbed methane (CBM) and 3 tonnes per cubic metre of tight gas. However, the place that gas holds in

China's primary energy consumption still remains marginal. In 2013, China's energy mix was 65.7 per cent coal, 17.8 per cent oil, 9.6 per cent non-fossil fuels and 5.1 per cent natural gas (BP Statistical Review 2014). In order to increase the gas component, the 12th Five-Year Plan established development objectives whereby its level would reach 8.3 per cent by 2015, 10 per cent by 2020 and 30 per cent by 2035, while the demand for gas would reach 230 billion cubic metres by 2015, approximately twice the total gas consumption registered in 2011.

The importance of gas has also been underlined by official speeches, in particular with a view to switching to low-carbon energy sources. At the Central Economic Work Conference held in Beijing in December 2013, the Chinese Premier Li Keqiang asserted the importance of adjusting the current structure of energy consumption (Central Economic Work Conference 2013). This restructuration would encompass an increase in the proportion of clean energy sources, insisting on the development of distributed energy resources, combining solar, wind and gas resources in decentralised cooling, heating and power systems (CPC Central Committee 2013), a view that essentially contradicts the current patterns of energy production and distribution.

Production and consumption records

At the beginning of the twenty-first century, China began to work actively on the exploitation of its gas resources. In 2000, national production was estimated at 27 billion cubic metres (EIA 2014), and it continued to witness an average yearly increase of 13 per cent until reaching 107.5 billion cubic metres in 2012 (ibid.).[3] Since then, the Chinese natural gas market entered a period of rapid development, with multiple natural gas sources gradually taking shape, and gas exploitation infrastructures being progressively developed. According to the US Energy Information Administration (EIA) (ibid.), gas production increased more than threefold between 2002 and 2012. Gas exploitations are mainly concentrated in three major production basins, namely the Inner Mongolia's Ordos basin, the Sichuan basin and the north-western areas of Xinjiang and Qinghai, which harbour the Tarim, Junggar and Qaidam basins. Concerning offshore resources, 142 gas fields were detected and distributed in the Bohai Gulf, the Pearl River delta and in some disputed zones of the South China Sea.

During the same period, the natural gas transportation and distribution system started to develop at a rapid pace, while the security level of the natural gas supply was on a constant increase. From 2000 to 2012, the average annual growth of the length of the Chinese long-distance gas pipeline remained at 13 per cent (NEA 2012). By the end of 2013, the length of the main artery and sub-artery reached 51,499 km, while construction plans anticipate reaching an extension of 250,000 km by 2015 (Newman 2014). The main part of the national pipeline network is broken down into five main lines. The longest and most significant one is the southwest pipeline, connecting southern Sichuanese resources with northern consumption areas. It has a total length of 5,923 km and is connected to the Zhongwu pipeline, the shortest in the national system, joining Hunan with Hubei province with a total length of 719 km and an annual capacity of 3.0 billion cubic metres. The West–East pipeline, currently divided into two branches, No. 1 and No. 2, runs from Xinjiang to Shanghai for a total length of 3,836 km and has a yearly capacity of 12 billion cubic metres. Finally, there is the Seninglan pipeline, which runs from Qinghai to Gansu province and has a total length of 931 km and a designated annual capacity of 3.4 billion cubic metres, and the Shaanxi–Beijing pipeline, of 910 km and with a capacity of 3.3 billion cubic metres/year[4] (see Fig. 3.1 for main pipeline network). As for city distribution pipeline networks, the overall length currently exceeds 460,000 km, with an average annual growth of 10 per cent (MoHURD 2012). Gas is carried in all 31 provincial and regional capital cities, with the last pipeline constructed to serve the city of Lhasa, Tibet (ibid.). As for LNG terminals, by the end of 2013, China already had nine major installations with a total capacity of 42.45 billion cubic metres (EIA 2014). Five other terminals are currently under construction, while the capacity for regasification is expected to increase by another 56.6 billion cubic metres by 2016 (ibid.). Underground natural gas storage is also under construction. China disposes of about 20 underground storage infrastructures that are mainly used to respond to natural gas consumption peaks in medium-sized and large cities or regions.

Together with the development of gas production and gas distribution facilities, Chinese natural gas consumption also witnessed rapid development between 2000 and 2012, passing from 24.5 billion cubic metres to 147.16 billion cubic metres in 2012 (ibid.), recording an average annual growth of up to 16 per cent. Although consumption is spread all over the country, the Eastern regions of the Yangtze River Delta, Circum-Bohai Sea, and southeast coastal areas absorb over 50 per cent of the total national consumption (MoHURD 2012), thereby highlighting an important divide between production areas and consumption areas, with understandable costs involved for the provision of gas infrastructure.

With regard to how consumption is structured, it has in recent years become increasingly diversified, with a predominant increase of gas usage in urban areas. From 2000 to 2012, the average annual growth rate of natural gas consumption in cities reached 29 per cent, three times that of natural gas production (NEA 2012). To get an idea in numbers, city natural gas consumption went from 2.94 billion cubic metres to 78.5 billion cubic metres in 12 years,[5] at the same time as there was a diversification of usage. Natural gas in cities played an increasingly important role in power generation, in replacing coal for heating and transportation, and in the development of distributed energy systems.[6] Following China's policies to solve the pressing problem of air pollution, currently at the heart of a real environmental crisis in major Chinese cities, these quantities are very likely to increase.

Figure 3.1 China's main pipelines, LNG import terminals and planned construction
projects
Source: Beijing Gas Company.

The switch to cleaner energy sources is encouraged by the Chinese government, wishing an 'intensive, intelligent, green and low-carbon new type of
urbanisation' (Central Economic Work Conference 2013), with gas to play an
important role. According to the President of the State Council's Development Research Centre (DRC), Li Wei, energy consumption will increase
by 60,000 tons of standard coal for each percentage point increase of urbanisation. In terms of natural gas, Beijing Gas Group Research Centre estimates needs to be around 15 billion cubic metres. Moreover, according to
MoHURD (2012) estimations, the rate of popularisation of natural gas
(calculated by dividing the total Chinese population by the total number of
gas users) will increase from an average rate of 29 per cent in 2011 to 46
per cent by 2015.

An increasing demand pushed by new energy and environmental policies

As coal combustion was the main culprit of air pollution emergencies
recorded in several cities in China between the winters of 2011 and 2013,

the Chinese government decided to respond with a series of new policies that limit the use of coal for power generation, replacing it with gas. In September 2013 in particular, the Chinese government issued an *Action Plan for Air Pollution Prevention and Control*, setting a roadmap for the following five years to reduce air pollution in three main regions, the Beijing-Tianjin-Hebei region, the Yangtze River Delta and the Pearl River Delta regions, all striving to achieve a negative growth of coal consumption. According to this document, the natural gas pipelines transportation capacity should increase by over 150 billion cubic metres by 2015 in the regions most affected by air pollution (State Council 2013a). The government is also encouraging an efficient use of gas by promoting projects such as the construction of energy facilities that distribute natural gas.

With the aim of significantly improving the quality of air in the capital city, the city of Beijing also issued its own *Clean Air Action Plan* (2013–2017) (Beijing Municipal Government 2013) in the same month of September 2013. The document asserted that measures had to be taken towards improving the quality of air, such as reducing coal consumption, promoting clean energy use (including gas) and limiting highly polluting production capacities. By implementing such measures, the city plans to close 1,200 small polluting mills in the building materials, chemical, founding and furniture sectors by 2016 (Xinhua 2013a). On the one hand, the consumption of gas in the city is highly encouraged as part of a general promotion of cleaner energy strategies and a cleaner urbanisation process, on the other, the urgency to act against unbearable levels of air pollution makes gas one of the solutions to environmental emergencies that cannot be ignored. This situation will significantly affect future trends in gas consumption.

Policies to promote the supply: developing unconventional gas

In terms of production, the government has issued a series of new policies to promote the development and exploitation of natural gas, including unconventional gas sources. In order to make use of the developmental potential of these resources, the Chinese government decided to make exploration activities available to private companies (differently from conventional natural gas, where exploration and exploitation can only be operated by important national oil companies). For upstream industries especially, the promotion of the exploration and development of shale gas and CBM has been encouraged by the release of the *12th Five-Year Plan for Shale Gas Development and the 12th Five-Year Plan for CBM*, providing great support in aspects such as mining rights approval, foreign cooperation mode, and technical R&D.

The 12th Five-Year Plan for Shale Gas Development announced a production objective of 6,5 billion cubic metres by 2015. By 2020, the exploitation of this resource should reach one-third of the total national production of natural gas, namely 60–100 billion cubic metres. Predominantly concentrated in the southern provinces of China, such as Sichuan, Yunnan and the Chongqing region, the shale gas reserves are the subject of different estimations by the major international and national energy institutions. According to the US Energy Information Administration, China's shale gas resources amount to 36.09 tonnes per cubic metre, a number clearly above the estimations of the IEA (25.99 tonnes per cubic metre) and the Chinese Ministry of Land and Resources (25.08 tonnes per cubic metre). As for Petrochina, shale gas reserves in China amount to 30.69 tonnes per cubic metre.[7] To date, the Ministry of Land and Resources has identified 180 areas with shale gas reserves and classified them as zones for development priority (Fei 2012a). In December 2011, the Ministry also issued a notice in which the State Council approved shale gas as an independent mineral, a vital decision for the formation and development of the shale gas industry. Being listed as an 'independent mineral' actually means that there is no state-owned energy enterprise monopoly over this resource, thus potentially being explored and exploited by other companies. At the end of 2013, in order to support the development of this industry, the National Energy Administration (NEA) also issued a document entitled The Shale Gas Industrial Policy, representing the very first national policy for this resource. The document focuses on five important components for the development of the shale gas industry in particular: (1) the designation of shale gas as one of the nation's strategically emerging industries; (2) the provision of subsidies for shale gas producers; (3) the encouragement of provincial governments to subsidise local producers; (4) the definition of tax reductions or exemptions for producers; and (5) the establishment of customs tariff exemptions for imported equipment (Xinhua 2013b). The policy also wishes Chinese companies to use China-owned brands for shale gas exploration and extraction technologies, encouraging innovation and development.

In order to underline the importance of shale gas in China's energy strategy and its willingness to open up to 'market forces', the Chinese government also pushed for an acceleration in prospection works by launching two tenders for shale gas blocks, both in 2011 and 2012. The second tender in particular was opened to private companies, with a wish to start a 'shale gas revolution' similar to the US model, mainly based on the entrepreneurship of small private exploration companies. The first round included seven exploration blocks mainly concentrated in Inner Mongolia. Given that this round was open to state-owned companies only, the seven blocks were mostly adjudged to Sinopec, with the sole exception being a provincial company called the Henan Provincial Coal Seam Gas Development and Utilisation Co. During the second round of tenders, both private companies and joint ventures under Chinese control were encouraged to participate. The auction concerned 19 blocks that are scattered across eight provinces. As a result of this 'opening up', 91 enterprises tendered, of which 53 were public enterprises, five were joint ventures and 34 were private enterprises. Notwithstanding the strong emphasis in the official discourse regarding the opportunities for private companies to significantly contribute to shale gas exploration, this second round saw only two private companies winning block exploration rights. The blocks were finally won by a majority of local companies supported by local governments, as well as local investment groups in the energy sector. Two state-owned companies also got exploration opportunities, although they were specialised in the exploitation of the coal sector. The other successful companies lacked any expertise in shale gas development, therefore raising doubts about their capacities to genuinely contribute to the production objectives established by Chinese authorities. These doubts were quickly confirmed one year later, when enthusiasm for shale gas development gradually subsided due to the lack of substantial progress. Of the 16 companies that won the 19 blocks, only one block had been drilled by mid-2013. In spring 2014, it was recorded that only two companies had been able to make any progress, namely the country's two most important energy players, Petrochina and Sinopec, while only 100 wells had been drilled (compared to 40,000 in the USA). At the time of writing, the ambitious production target of 60 billion cubic metres by 2020 has been slashed by half (Reuters 2014).

With regard to CBM, the Chinese government established a development objective of 16 billion cubic metres by 2015 and 40 billion cubic metres by 2020. CBM volumes are mainly concentrated in the north and northeastern parts of the country, together with Sichuan and Xinjiang basins. As for shale gas, beyond the publication of a 12th Five-Year dedicated plan, the Ministry of Land and Resources also issued a CBM Industrial Policy, requiring 'scientific and efficient development and utilisation' of CBM resources. In September 2013, the State Council, willing to push for investments in CBM exploration and development, as well as for the construction of more transportation infrastructure, also issued a policy guideline. The policy foresees the adoption of financial incentives and tax breaks to encourage producers, as well as a reform of local price controls. However, as in the case of shale gas, the CBM industry is still in its early stage of development, sharing similar challenges with other unconventional resources (although its exploitation is easier from a technological point of view). As the EIA (2014) reports, the production of CBM is increasing in China, yet there are still many obstacles to overcome (from regulatory to technical and infrastructural aspects, especially the lack of pipelines from coal mining areas to gas markets), hindering the full exploitation of its potential. Not exclusive to CBM or shale gas, most of these problems have to be considered as the main obstacles to further development in the natural gas sector.

Technical and institutional challenges to gas sector development

Notwithstanding the richness of resources and with an increasing demand for natural gas, encouraged also by the government's recent 'cleaning up' policies, the gap between supply and demand is likely to be on a constant increase in the coming years. The gap recorded was around 40 billion cubic metres already in 2012 (EIA 2014). Moreover, the Chinese government's objectives are not likely to prioritise self-sufficiency. According to the *12th Five-Year Plan for Energy Development*, national production should reach 156.5 billion cubic metres by 2015, while demand will reach 230 billion cubic metres (State Council 2013b). EIA projections are less optimistic. According to the US agency, Chinese gas production will only attain 118.93 billion cubic metres by 2020, while there will be a demand for 220.87 billion cubic metres. However, no matter which numbers are more convincing, EIA projections and the 12th Five-Year Plan for Energy Development both underline a growing gap between supply and demand. This increasing difference can be ascribed to management shortcomings, namely insufficiently developed infrastructure, an obsolete price mechanism and an institutional organisation of the sector, hindering the potential for the marketisation of the sector.

A problematic lack of infrastructure

An important factor that impedes the possibility for national supplies to match domestic demand is the lack of a sufficient level of infrastructure capable of carrying gas from the centres of production, primarily situated in the north, to the major consumption areas, namely cities in the east. To get an idea of the scope of this problem, the total extension of the pipeline network was nearly 51,500 km in 2013, representing just one-tenth of the total pipeline network of the USA. As a consequence, the limited transportation capacities currently available do not encourage the development of new production of gas. Moreover, for a long time and until very recently, monopolistic actors prevented the access to existing infrastructure, describing a physical and 'institutional' insufficiency of transportation capacities.

First of all, there are objective difficulties that China faces if it wishes to develop transportation and distribution infrastructure that is flexible enough to manage seasonal consumption peaks. For example, in Beijing, high consumption peaks can reach 60 million cubic metres per day, while in periods of low consumption demand can drop to a daily 5 million cubic metres. Thus, there is an important need to develop an infrastructure that is capable of managing such steep changes in consumption, avoiding the gas shortages recorded in the winters of 2004, 2009 and, more recently, of 2013. Beyond the difficulty of providing infrastructure to respond to a highly variable demand, there is also the problem of available land in which to introduce the pipelines. Perhaps what is most important in the Chinese economy is land, increasingly rare for agricultural exploitation and under the greedy attention of local governments to developing real estate and income generation (see Romano, Chapter 11). This is particularly problematic for the construction of vehicle fuel stations.

To respond to the limited availability of transportation infrastructure, the government fixed important objectives for the development of a pipeline network, which is likely to reach 250,000 km by 2015. However, meeting this objective is strictly connected to the ability of the sector's major enterprises to invest in new constructions. In recent years, China's three main oil and gas companies – Sinopec, CNOOC and Petrochina, which monopolise the upstream activities of the sector – have witnessed an important reduction in their profits and increase in their debt. For instance Petrochina, the primary gas producer nationally (covering 70 per cent of total gas production in China), saw its debt to capital ratio reach 47 per cent in June 2013 (Daiss 2013). At the heart of this poor performance are the costly gas imports, not compensated for by an adequate level of gas prices. Petrochina, which imports two-thirds of China's total gas imports, buys gas at international market prices and sells it at the prices fixed by the National Development and Reform Commission (NDRC), which are voluntarily kept low to protect consumers. This means that the national company registers an average loss of 1 RMB for each cubic metre of gas sold. In 2013, Petrochina's total loss reached 49 billion RMB, of which 28.3 billion were due to the imports of 27.45 billion cubic metres from Central Asia, and 420 million RMB to the 409 million cubic metres from Myanmar. The average annual loss was 1.03 RMB per cubic metres (Platts 2014a). As for LNG imports, Petrochina registered a loss of 20.3 billion RMB for the sale of 7.34 billion cubic metres of imported LNG, with an average loss for LNG of 2.77 RMB per cubic metres (ibid.). If these numbers stagnate, the company's debt ratio could even break 50 per cent by 2015 (Daiss 2013). With this in mind, the possibilities for financing important infrastructure investments are quite limited, with important consequences arising for national gas production objectives. In the case of CBM – very foreseeable also for shale gas exploitation – the lack of infrastructure rendered the conveyor in the transportation network of the gas produced difficult. In 2012, a mere 41 per cent of the gas produced was used, while the rest was burned. For a country with a rising energy demand and an increasing need for cleaner energy sources, such a waste can hardly be considered as part of a rational energy security strategy.

Technical challenges can also be observed in exploration and exploitation, namely in the unconventional gas sector. The geological complexity of shale gas reservoirs makes the possibility to exploit these resources intrinsically linked to the availability of sophisticated technologies, which are currently lacking. As previously observed, China had ambitious development objectives for shale gas. These were eventually halved. According to experts at a conference that centred on the place of gas in China's energy security strategy organised by Asia Centre in June 2013, without significant technological improvements the production objective for shale gas development is very likely to meet the same end as CBM: a failure. As seen in the 12th Five-Year Plan, the production objective for CBM is 16 billion cubic metres, but production estimates can only confirm a possible 6 billion cubic metres. Already in 2010, the press pointed out the need for revising objectives through a 'reality check' (Hook 2010). At that time, China wished to reach an annual 5 billion cubic metres, but the reality showed just a quarter of that. The lack of transportation infrastructure is attributed to the disappointing results of the sector, as third-party pipeline access had been limited for a long time (ibid.).

To face these technological obstacles, China is investing on a large scale in the development of new technologies, all the while promoting partnerships with foreign companies in order to explore and exploit both national and international gas resources. For instance, in March 2012, Petrochina signed the first shale gas production-sharing contract with Royal Dutch Shell with a view to the joint exploration and development of the rich Fushun-Yongchuan block in the Sichuan basin. In February 2013, CNOOC succeeded in acquiring the Canadian company Nexen, providing the Chinese enterprise with the opportunity to learn from Nexen's shale gas exploitation of British Columbia. However, the complexity of Chinese geology that we have highlighted makes existing foreign technologies unfit for exploration. Coupled with the problem of the environmental and geological impacts of the use of hydraulic fracking as a means to exploit shale gas resources, there are important obstacles to the possibility of triggering a 'shale gas revolution' in China. In a country where a shortage of water resources is becoming an increasing problem, hydraulic fracking may look like the most unsuitable option, not only because considerable quantities of water are needed, but also because chemical products are employed during the exploration process, and there are important dangers of contaminating groundwater resources if these products are not adequately managed. A mismanagement of shale gas wells could also lead to greenhouse gas leakages, as well as pollutants such as PM 2.5, not to mention noise pollution. Moreover, since shale gas fields are mainly concentrated in highly seismic areas, their exploitation seems to be unfit for the use of hydraulic fracking. Shale gas exploration also has a significant land use imprint, returning to the issue previously highlighted regarding the construction of new gas pipelines capacities.

Beyond the technical and environmental problems, institutional barriers also constitute important hurdles along the road to enhancing natural gas production at a national level, either in exploitation or in transportation. Compared to the framework that permitted the USA to have its 'shale gas revolution', China currently lacks the conditions necessary to trigger entrepreneurship of private companies. In the USA, the considerable development of the shale gas industry is due to the presence of an open market, few entry barriers, the entrepreneurship and competition of small companies (pushing them to improve their technological capacities), as well as encouraging fiscal measures (low taxes) and effective regulations. In other terms, the USA developed a favourable environment for the exploitation of potentials incurred by the competition of small companies and market mechanisms. Moreover, a well-developed system of transportation infrastructure (the USA disposes of about 500,000 km of pipelines) and clear regulations for its access provided the necessary confidence for shale gas producers to invest in the exploitation of resources. Considering China's current situation, these conditions are clearly lacking. Market entry barriers are quite high, while infrastructure access to private and small companies is complicated. Furthermore, regulations are not complete unless all the aspects of shale gas production and transportation have been taken into account, from exploration to pipelines access, environmental protection, etc. Obviously these issues have a solution, and the National Energy Administration is currently working on the preparation of specific regulations for shale gas exploration and exploitation, as its representatives explained during the conference on gas in June 2013. Technological problems can also be solved through international cooperation and further investments in R&D. Moreover, the access to private companies can be further enhanced, provided that more competent players participate in future auctions. However, there is still an important problem that is discouraging the investment in and development of shale gas, as well as the enhancement of gas transportation capacities, and that is low gas prices.

Gas prices: at the core of the energy security problem

Regardless of whether the issue is taken from a national production point of view, or whether it considers international trade relations, energy security in the gas sector is intimately linked to the question of gas prices reform. Observing a strict control over energy prices currently discourages both the private and the public sector to invest in shale gas development, a resource that requires a high and continuous level of investment for its maintenance. When it comes to shale gas, drilling is quite an expensive operation and it also demands a certain frequency. As wells are decreasingly productive, companies are obliged to drill several in order to maintain production. The lack of infrastructure and the limited access to what currently exists further inhibits potential investors, while few public resources are available for investing in new infrastructure. Only until very recently was third-party access to pipeline network not guaranteed. Giant oil companies used to control upstream gas exploration, as well as pipeline construction and delivery. However, in order to break with their virtual monopoly and further boost the access to new players in gas production, as well as tapping the potential for a progressive marketisation of the energy sector, the Chinese government issued guidelines allowing third-party access to crude, oil products and gas pipelines in February 2014 (Platts 2014b). This was presented as a way for the government to encourage the unconventional gas sector by helping other players gain access to the transportation network.[8] However, notwithstanding this move, there is still a great need for new projects in infrastructure, which demand larger efforts from already indebted oil and gas companies. Furthermore, if private companies are required to participate in the development of shale gas or other unconventional sources, then they need to be sure that their investments are matched by subsequent profits from gas sales. With this is mind, a reform in gas prices is ineluctable.

Signs of a reform...that requires rapid implementation

In China, natural gas prices are established based on three different components: ex-plant prices, pipeline transportation fees and city gas distribution fees. These prices are set by the NDRC. In the case of the first kind, suppliers and purchasers agree on a purchase and sale price that is no higher than 10 per cent over the guide price established by the NDRC. This price is adjusted on a yearly basis and is based on the average price over the previous five years of crude oil (weighing for 40 per cent), of liquefied petrol gas (20 per cent) and of coal (40 per cent), and includes a limit for adjustments equal to 8 per cent (Fei 2012b). In the second case, pipeline fees are established according to specific pipeline projects. The price can be set by the NDRC or it can be agreed on following a guide price (ibid.). Finally, city gas prices result from the sum of the first two types of prices and are subject to strict government control. Prices at the city level are then differentiated according to the type of user (residential, industrial or commercial), while the city government and the provincial pricing bureaus establish distribution fees (ibid.). As for unconventional gas and LNG, prices are negotiated according to the international market prices established by gas producers and their clients, which could push international companies to invest in the exploration and exploitation of shale gas.

In order to encourage investments in the sector, the NDRC proposed to link natural gas prices to international oil prices, consequently increasing gas prices for end users and laying the grounds for a deeper reform in gas prices. At the end of 2011, a new pilot scheme was introduced in two provinces, Guangdong and Guangxi. The scheme involved regulation of the city gate price of natural gas according to a hub price based in Shanghai. Previously, the city gas price had been based on gas usage or on the specific ex plant price. Through the reform, establishing a gas price became more market-oriented, following the prices of imported fuel oil and liquefied petroleum gas (LPG). Given the initial success of this experiment, the NDRC then proceeded to a first gradual diffusion to the rest of the country. In July 2013, it introduced a two-level price setting, where the first level was based on the national gas consumption of 2012, and the second level was pegged to international LPG and oil prices. Within the total quantity of 112 billion cubic metres, the city gate price is negotiated by the interested parts, yet can only vary within a price range fixed by the government for each province (approx. 0.4 RMB per cubic metre). Beyond this quantity, prices for extra gas consumption for 2013 are linked to international LPG and oil prices, with a 60 per cent weight for oil price and a 40 per cent weight for LPG price, followed by a 15 per cent discount to reflect the price advantage of natural gas. Through this reform, the NDRC has thus increased non-residential gas prices from 1.69 RMB cubic metres to 1.95 RMB, for a total 15.4 per cent increase. The average differential between the two pricing bands is 40 per cent. In its work report of March 2014, the

NDRC announced to the annual session of the National People's Congress that gas prices would be further increased later in the year.

Although these recent reforms in pricing mechanisms have proved to be encouraging and have promised further improvement in the gas prices reform, there are still important shortcomings that cannot be neglected. First of all, regarding the quantities of gas interested by the price reform, 90 per cent of the gas distributed belongs to the first kind, meaning that a mere 10 per cent is sold according to the international market price level. Moreover, as the experts who participated in the June 2013 conference highlighted, the chemical fertiliser industry, which is an important gas consumer, still remains highly subsidised by the government, with the possibility of buying gas with a limited price increase of 0.25 RMB per cubic metres. Finally, residential users, who account for 20 per cent of total consumption, have been clearly untouched by the reform. The NDRC stated that gas prices for this category of users would also gradually be modified, while hoping to complete the implementation of the price reform by 2015 when gas prices will be aligned with international ones. However, the different players involved in the gas sector (from big companies to local distributors and consumers – or better yet, the potential reaction of consumers) show diverging interests, which hinder progress in negotiations and significantly delay the much needed price reform.

Diverging interests and underlying contradictions

One reason for China to hesitate in introducing market mechanisms to determine gas prices is the potential reaction of residential consumers. In the first version of the price reform, the Chinese government carefully avoided addressing the residential sector, fearful of popular reactions. This reason was presented several times during the conference held in June 2013, where experts highlighted the importance for reformers to carefully guarantee a correct price for individuals. At other meetings, experts also expressed the difficulty for the government to tackle prices, fearing possible reactions of *laobaixing* (literally 'the old 100 family names', an expression that refers to Chinese citizens). Beyond the possible reaction of residential users and a general reduction in the demand for gas, another reason why a price reform is not desirable is its probable contradiction with the policies adopted by local governments to replace coal. Since gas is still an expensive option, it must be subsidised in order for it to find opportunities to replace coal. However, cash-strapped local governments – obliged to send land use rights to guarantee a source of revenue in order to finance their infrastructural projects – already find it very difficult to subsidise the use of gas. A further increase in prices would definitely put an end to local experiments that help gas in replacing coal, based on specific policies of encouragement. As long as coal prices remain low and do not include the costs of its externalities – through carbon tax, for instance – it is highly likely that gas (and other clean resources) will hardly find a place in a city's choice of energy, contradicting the central government's wishes for cleaner energy sources.

The presence of differentiated prices according to the type of consumer, as well as price differentials for provinces, also show the government's difficulty in satisfying all the interested parties. While only a few players have a monopoly over the upstream sector, a considerable fragmentation can be seen in the downstream section, requiring the government to carefully dose its reform steps so that the final consumers are not the only bearers of price increase. However, behind this discourse – understandable in a country where a considerable part of the population is still unable to afford international prices in energy – there are also important and seemingly contradictory interests in keeping the status quo. Once again, the answer should be found on the big NOCs side. Although their sales of foreign gas are made under non-profitability conditions, these companies block the entrance to other actors within the gas sector. A liberalisation of gas prices should pass through a development of market competition, which implies letting third parties enter gas production and transportation infrastructure – thus partially losing their control and monopoly. Under such uncertain conditions and a sluggish gas price reform, it is understandable that foreign companies refrain from taking great risks in participating in prospection activities, whereas their presence would be very important for unconventional gas exploration. If foreign companies entered, it would significantly help China to trigger the conditions needed to exploit its difficult basins, as these companies not only have greater possibilities to develop and introduce the technologies and know-how that are currently lacking in China, but they also invest in this highly demanding extraction activity, therefore pouring necessary financial resources into China. Once again, a full participation of foreign and private companies in exploration and production activities is perhaps something that the Chinese government – or better yet, the branches administering the big state-owned energy companies – is still not ready to let happen. In these conditions, the only way to match the increasing demand is by increasing imports. However, this implies energy dependency and security repercussions.

Increasing insecurity: cleaner energy means higher dependency

Even if China establishes policies and regulatory frameworks to push for an improvement in national gas production capacities (by significantly assisting in shale gas development) within the next few years, it is unlikely that these efforts can promptly cover the existing and ever-widening gap between supply and demand. In other words, for China to replace coal and pursue its clean energy strategy, it will still need to import gas from foreign countries, increasing its dependency from foreign sources as well as the risks of supply disruptions. When it comes to considering China's foreign resources provision, several energy security questions emerge, and although this chapter does not focus on international relations in energy, the gas perspective forces us to make considerations on some aspects. Traditionally a net gas exporter, China became a net gas importer for the first time in 2007. Between 2010 and 2011, total dry natural gas imports almost doubled, from 16.34 billion cubic metres to 31.37 billion cubic metres,[9] finally reaching 53 billion cubic metres in 2013, thus representing one-third of China's gas consumption (Xinhua 2014). In only seven years, China's gas–foreign trade dependence ratio reached 29 per cent (EIA 2014). Between 2007 and 2012, imported gas from pipelines went from 4.4 billion cubic metres to 22.8 billion cubic metres, with an average annual growth of up to 32 per cent, while LNG import passed from 3.9 billion cubic metres to 20 billion cubic metres, with an average annual growth of up to 39 per cent.[10]

At present, China's major source of gas import is Central Asia, covering more than half of the country's imports. As Fig. 3.2 shows, the bulk of provision comes from Turkmenistan, which supplied China with 51.4 per cent of its gas imports in 2012. Then follow Qatar (16.4 per cent), Australia (11.7 per cent), Indonesia (8 per cent) and Malaysia (6.1 per cent), with the rest of China's providers delivering between 2 per cent and less than 1 per cent of gas resources. Although China tries to not to discriminate among its gas providers, the heavy reliance on Turkmen gas, and in smaller measure on Qatari gas, raises questions concerning the security of these areas, as they belong to regions under threat of possible terrorist attacks, as well as ethnic or religious conflicts. These risks add to other more common dilemmas that weigh on energy import, such as financial and economic risks, reserve change issues, infrastructure damages, water and land availability, etc.

Notwithstanding China's efforts to develop long and stable relationships with Central Asian countries, as well as its ability to obtain favourable conditions for gas provision – having signed a contract of participation in exploration and exploitation of its gas fields with Turkmenistan (see Boas and Spaiser, this volume, Chapter 7) – the risks of over-relying on the resources of this area are very high, inevitably requiring China to diversify its sources. Recent tensions between Kyrgyzstan and Tajikistan in March 2014 – exchange of gunfire at the border between the two countries, escalating since the beginning of 2014 – and the closure of their border led to important disruptions in traffic and trade (Radio Free Asia 2014). China's plans to build a new Central Asian line (the so-called 'line D') beginning in Turkmenistan and passing through the Kyrgyzstan and Tajikistan borders surely entails a great component of risk. Other risks are also connected to the nature of these regimes, characterised by 'low investments, corruption and gross mismanagement' (International Crisis Group 2007).

In terms of cost – since piped gas remains cheaper than LNG – the regional option is still preferable, which would also render Russia a perfect candidate for China's gas provision. However, long-term negotiations with Russia on gas sale prices – finally coming to a successful conclusion with the last visit of the Russian President to China in May 2014 (Hornby *et al.* 2014) – showed that gas prices (and the cost of infrastructure development)

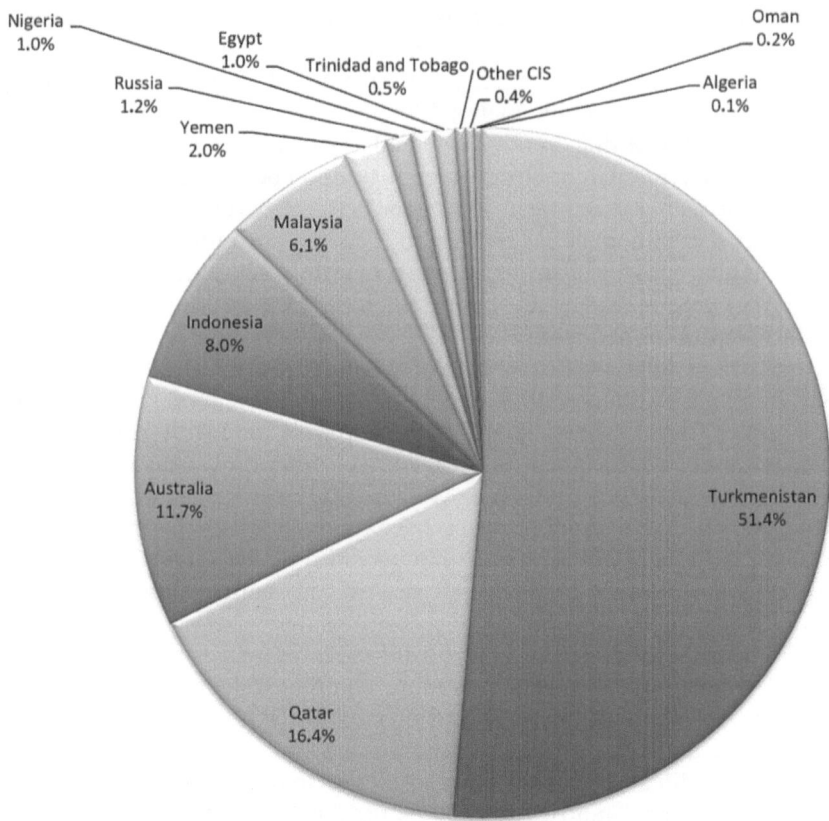

Figure 3.2 China's gas imports per country
Source: Re-adapted from BP Statistical Review of World Energy.

are an important factor that China needs to consider carefully, notwith-standing how close the two countries are and the 'mutual interests' they have in engaging in a gas deal. An article in the *Diplomat* clearly points out this 'mutual interests' vision and how it is not matched with the reality of the situation. From a theoretical point of view, the gas relationship between China and Russia can be characterised by complementarity: Russia is the world's richest country in natural gas reserves, with its major gas fields concentrated in Siberia and in its Far East. The road to piping gas to China is thus shorter than its gas provisions to Europe. However, notwith-standing this 'ideal' situation, the pace of gas negotiations has been slug-gish. 'Various technical obstacles, pricing conflicts and mutual suspicions have historically kept Chinese purchases of Russian energy at relatively low levels' (Weitz 2011). Moreover, '[f]requent delays in shipments on the part of the Russians and attempts to leverage the competing interests of the Chinese,

Asian, and European markets off of each other have prevented Chinese policy makers from regarding Russia as a reliable long-term supplier' (ibid.).

Eight years of negotiations between China and Russia have showed that these 'mutual interests' in increasing bilateral energy cooperation are far from being a given. Notwithstanding the recent improvement in the form of a gas deal, there are still important details to be discussed, like the final gas sale price. According to the media, the implied price of the negotiation was US $350–390 per 1,000 cubic metres, very similar to what European countries pay to get Russian gas. However, further negotiations towards setting a final price were expected for the end of 2014, as the current final price has long been a bone of contention. Once again, the November 2014 deal did not lead to an agreement regarding the sale price of gas (O'Sullivan 2014). China negotiated hard for Russia to invest US$55 billion in the development of the gas fields of Chayanda and Kovykta in Eastern Siberia, as well as in the pipeline to the Chinese borders. China will provide an investment of US$20 billion, although it is still unclear what the money will go towards. Interestingly, the development of these gas fields will also feed the designed LNG plant of Vladivostok, which will provide other 'hungry' East Asian countries, such as Japan and South Korea (as well as putting the three north-east Asian countries in competition for gas supply from Russia). The development of these gas fields and the pipeline construction have to be seen not only as an important progress in Sino–Russian energy relations, but also as a negotiation advantage for China in terms of gas supply diversification and import prices. As a result of the pipeline, gas from Russia will be paid at a lower price, which will give China more leverage in negotiating prices for LNG imports (Tuttle *et al.* 2014). In 2012, most LNG imports came from Qatar (34 per cent), followed by Australia (24 per cent), Indonesia (16 per cent), Malaysia (13 per cent) and Yemen (4 per cent), to cite the most important suppliers (EIA 2014). In 2012, the average LNG import price for all terminals was US$10.43 per million British thermal unit, although this does not reflect the variety of prices paid by different countries for LNG imports. For instance, in April 2014, China paid Australia US$3.25 per million British thermal unit, whereas it paid US$19.97 per million British thermal unit (ibid.) in November 2013 for gas coming from Algeria. Moreover, in some terminals, such as that of Jiangsu or in Dalian, the LNG price was around US$17 per million British thermal unit, which already reflects the high gas prices paid in Northeast Asia.[11] Thanks to the recent Russian agreement, LNG-supplying countries will be forced to offer more competitive prices if they want to ink contracts with China. As the country has been progressively diversifying its gas resources by signing agreements with different gas suppliers (above all Turkmenistan), it is becoming harder for suppliers like Qatar and Russia, or other possible players, to impose high prices on China. In March 2013, the conclusion of an agreement for a 20 per cent participation in an ENI's offshore basin in Mozambique showed that China will try its best to have the maximum leverage in negotiating low supply prices – and that buying stakes in production is the preferred option, thus having the possibility to control prices. Signing agreements with Australia – likely to become the world's largest LNG supplier by 2020, surpassing Qatar – is giving further leverage to China. Moreover, since 2009, China has been participating in the construction of an oil and gas pipeline from Myanmar, which will enable China to be supplied by that country's gas. Also in 2009, the launch of the vast pipeline extension project connecting Turkmenistan, Uzbekistan, Kazakhstan, China and perhaps even as far as Iran, confirms that China is tightening its relations with these countries and will continue to support Iran's energy exports. Finally, the 'shale gas revolution' in North America is also currently offering other possibilities to China to invest in upstream developments. The most important achievement in these terms has been CNOOC's acquisition of a Canadian company, Nexen, which gave the Chinese company the chance to benefit not only from its fields, but also from its expertise.

Notwithstanding these developments, it is not yet sure how long China will be able to play this elusive strategy in dealing with its gas suppliers. The diversification strategy is surely a way for China to ensure not to pay exorbitant prices for the gas it imports, putting the different suppliers in competition with each other, at the same time that it is conscious of being a too important buyer for every gas producer to simply be dropped. But due to recent gas shortages in the winter of 2013, pushing the Chinese government to adopt a policy on limiting the gas supply to industries in order to guarantee household gas supply could already be a sign that the supply–demand gap will likely weaken the Chinese stance in waiting for gas suppliers to cede to its price requests. Only recently, the Chinese government revised its gas consumption projections by 2020, anticipating future consumption at 400–420 billion cubic metres (Downs 2014). In the meantime, unless China does not find other supply opportunities, either abroad or in its own territory, the possibilities of negotiating lower prices will progressively recede. The final price set for Russian gas purchases could be a sign whether China still has leverage or whether this is in fact progressively weakening.

Conclusion

Based on this overview of gas security in China, it is likely that the objectives anticipated by the 12th Five-Year Plan for gas development in the country's energy mix will not be met. According to experts who participated in Asia Centre's June 2013 roundtable, the Chinese government's forecasts of gas deficits are underestimated, as the gap between supply and demand is likely to be greater. Although the 12th Five-Year Plan predicts 54 billion cubic metres and 80 billion cubic metres as the gas deficit in 2015 and in 2020 respectively, experts highlighted how these numbers could be greater – and recent revisions of the Chinese government's forecast of future gas consumption by 2020 of up to 400–420 billion cubic metres confirm these doubts. China is currently facing a gas challenge that will weigh importantly on the country's wish to reduce its environmental impact by choosing cleaner energy resources. In the next few years, unless China is able to set a developing framework that is favourable enough to trigger a real 'shale gas revolution', the country's dependence on foreign sources will increase energy security concerns in all the respects that we have tried to encompass through this collective work. If energy security is not only the presence of reliable supplies – highly dependent on stable geopolitical relations and relatively stable prices for resources – but also encompasses other types of security, above all environmental and health security, then gas will be the reason for many of the Chinese government's headaches.

Beyond the need to diversify its gas sources and bargain hard for the best conditions in imported supplies, negotiations and balancing efforts should also be done among its domestic players, the interests of the various actors in the supply chain and, finally, of the end-users being differently affected by an improvement in price reform. The current configuration of the sector – with few oligopolistic actors in the upstream sector and many players in the downstream sector – does not contribute to the acceleration of reforms, which are vital for China to enhance its ability to switch to cleaner energy sources and significantly reduce the use of coal. Reforms are likely to take account of current power configurations, the positions of energy security in direct conflict with the irrationalities of sheltering behind a traditional energy stance where strategic resources should stay in the hands of small number of actors. Chinese NOCs will perhaps prefer to continue losing money while importing foreign gas instead of letting other actors nibble on their supply monopoly. Moreover, the presence of subsidised gas prices is likely to harden the implementation of energy saving policies, which are highly cherished by the Chinese central government (although such prices can still guarantee energy access to poor Chinese citizens). From an environmental and health security point of view, the current gas governance goes against energy security if this is to be seen as a stable, reliable and clean supply of energy resources. In total contradiction to the central government's policy discourses, this situation simply implies that coal will remain the main pillar of China's energy supply for a long while to come.

Notes

1 Most of this information comes from Asia Centre's research on gas in the year 2013–2014. In June 2013, we held a conference on the topic of gas in China's energy security strategy, in preparation for which it conducted several interviews and exchanges with experts, including representatives of both international and Chinese companies, Chinese ministries and NDRC representatives, scholars, think tanks, as well as representatives from embassies.
2 We have decided not to take this sector into account in this chapter, as it is addressed in Chapter 2.
3 The EIA's data are presented in billions or trillions of cubic feet. We decided to present the quantities in cubic metres.

4 Information about pipelines was retrieved from CNPC's official website, www.pet rochina.com.cn/ptr/yqssfw/commonlist_norig.shtml
5 Data from the Research Department of Beijing Gas Group.
6 Distributed energy systems combine cooling, heating and power supply systems through methods of heat and power cogeneration.
7 We compared the data available within the framework of our research on gas.
8 This 'optimism' should be nuanced in light of the current situation of the shale gas sector in terms of blocks possession. Despite the attribution of new shale gas exploration blocks through the two rounds of bids, the richest fields still remain in the hands of the big NOCs, about 80 per cent of all recoverable shale gas resources. These shale gas blocks are based in conventional gas fields, by law under NOCs sole exploitation. As for conventional gas, the Chinese shale gas industry is also characterised by an oligopolistic structure dominated by major national oil and gas companies. Despite the categorisation of shale gas as an independent mineral, most of its de facto exploitation remains in the hands of large state-owned companies.
9 We retrieved data from the US Energy Information Administration's International Energy Statistics, available at: www.eia.gov/cfapps/ipdbproject/IEDIndex3.cfm.
10 We retrieved data from the General Customs Administration of the People's Republic of China, available at: www.customs.gov.cn/publish/portal0/tab49667/.
11 This gas supply comes from Qatar, which makes its LNG exports prices tied to international oil prices.

Bibliography

Beijing Municipality Government (2013) *Beijing shi 2013–2017 nian guoji kongqi xingdong jihua [Beijing Clean Air Action Plan (2013–2017)]*. Beijing: Beijing Municipality Government. Available at: http://zhengwu.beijing.gov.cn/ghxx/qtgh/t1324558.htm (accessed 17 June 2014).
BP Statistical Review (2014) *China in 2013*. London: BP Statistical Review of World Energy. Available at: www.bp.com/en/global/corporate/about-bp/energy-economics/statistical-review-of-world-energy/country-and-regional-insights/country-insights-china.html (accessed 17 June 2014).
China Central Economic Working Conference (2013) *Premier Li Keqiang Speech from the Central Economic Working Conference of the People's Republic of China*. Beijing: China Central Economic Working Conference.
China CPC Central Committee (2013) *"Decision" of the CPC Central Committee on Deepening the Reform of Counseling Books*. Beijing: People's Republishing House.
China Ministry of Housing and Urban-Rural Development – MoHURD (2012) *Quanguo chengzhen ranqi fazhan shierwu guihua [12th Five-year Plan on City Gas Development]*. Beijing: China Ministry of Housing and Urban-Rural Development.
China National Energy Administration – NEA (2012) *Tianranqi fazhan shierwu guihua [12th Five-year Plan on Natural Gas Development]*. Beijing: China National Energy Administration.
China State Council (2013a) *Daqi wuran fangzhi xingdong guihua [Action Plan for Air Pollution Prevention and Control]*. Beijing: State Council. Available at: www.gov.cn/zwgk/2013-09/12/content_2486773.htm (accessed 17 June 2014).
China State Council (2013b) *Nengyuan fazhan shierwu guihua [12th Five-Year Plan for Energy Development]*. Beijing: State Council. Available at: www.gov.cn/zwgk/2013-01/23/content_2318554.htm (accessed 17 June 2014).

Daiss, T. (2013) 'PetroChina's Growing Debt Problem Projected to Worsen.' *Energy Tribune* (17 June). Available at: www.energytribune.com/77747/petrochinas-gro wing-debt-problem-projected-to-worsen#sthash.kQ67khc8.dpuf (accessed 17 June 2014).

Downs, E. (2014) 'In China-Russia Gas Deal, Why China Wins More.' *Fortune* (20 June). Available at: http://fortune.com/2014/06/20/in-china-russia-gas-deal-why-china -wins-more/ (accessed 15 November 2014).

Fei, K. (2012a) 'Shale Gas: A Game Changer for China's Energy Consumption Pattern?' *Norton Rose Fulbright* (March). Available at: www.nortonrosefulbright. com/knowledge/publications/64018/shale-gas-a-game-changer-for-chinas-energy-con sumption-pattern (accessed 17 June 2014).

Fei, K. (2012b) 'The Natural Gas Pricing System in China'. *Norton Rose Fulbright* (May). Available at: www.nortonrosefulbright.com/knowledge/publications/66881/ the-natural-gas-pricing-system-in-china#section6 (accessed 17 June 2014).

Hook, L. (2010) 'Doubts Over Chinese Coal-Bed Methane', *Financial Times*, 29 August. Available at: www.ft.com/intl/cms/s/0/cfd6258a-b38d-11df-81aa-00144fea bdc0.html#axzz3A0G3Iin3 (accessed 17 June 2014).

Hornby, L., Anderlini, J. and Chazan, G. (2014) 'China and Russia Sign $400bn Gas Deal', *Financial Times*, 21 May. Available at: www.ft.com/intl/cms/s/0/d9a 8b800-e09a-11e3-9534-00144feabdc0.html? (accessed 17 June 2014).

International Crisis Group (2007) 'Central Asia's Energy Risks.' *International Crisis Group* (24 May). Available at: www.crisisgroup.org/en/regions/asia/central-asia/ 133-central-asias-energy-risks.aspx (accessed 17 June 2014).

Newman, N. (2014) 'Reducing Market Barriers Seen as Key to China's Gas PipeDream.' *Pipeline and Gas Journal* 241(11). Available at: www.pipelinea ndgasjournal.com/reducing-market-barriers-seen-key-china%E2%80%99s-gas-pipe-d ream (accessed 29 November 2014).

O'Sullivan, M.L. (2014) 'New China-Russia Gas Pact is No Big Deal', *Bloomberg*, 14 November. Available at: www.bloombergview.com/articles/2014-11-14/new-china russia-gas-pact-is-no-big-deal (accessed 29 November 2014).

Platts (2014a) 'PetroChina's 2013 Losses on Imported Natural Gas Hit Roughly $8 Billion', *Platts*, 20 March. Available at: www.platts.com/latest-news/natural-gas/hon gkong/petrochinas-2013-losses-on-imported-natural-gas-21363390 (accessed 17 June 2014).

Platts (2014b) 'China to Open Oil, Gas Pipelines for Third-Party Access', *Platts*, 25 February. Available at: www.platts.com/latest-news/oil/singapore/china-to-op en-oil-gas-pipelines-for-third-party-27970194 (accessed 17 June 2014).

Radio Free Asia (2014) 'China Runs Risk with New Gas Route Through Central Asia.' *Radio Free Asia* (24 March). Available at: www.rfa.org/english/commenta ries/energy_watch/energy-03242014125135.html (accessed 17 June 2014).

Reuters (2014) 'China Finds Shale Gas Challenging, Halves 2020 Output Target', *Reuters*, 7 August. Available at: www.reuters.com/article/2014/08/07/us-china-sha le-target-idUSKBN0G71FX20140807? (accessed 29 November 2014).

Tuttle, R., Shiryaevskaia, A. and Almeida, I. (2014) 'Russia-China Deal to Damp LNG Prices as Output Rises', *Bloomberg*, 22 May. Available at: www.bloomberg. com/news/2014-05-21/russia-china-deal-seen-damping-lng-prices-as-outpt-rises.html (accessed 17 June 2014).

US Energy Information Administration (EIA) (2014) *China – Overview*. Washington, DC: US Energy Information Administration. Available at: www.eia.gov/countries/ cab.cfm?fips=ch (accessed 17 June 2014).

Xinhua (2013a) 'Beijing Unveils Clean Air Action Plan', *Chinadaily*, 12 September. Available at: http://usa.chinadaily.com.cn/business/2013-09/12/content_16966227. htm (accessed 17 June 2014).

Xinhua (2013b) 'China's Shale Gas Policy Promises More Financial Support', *People's Daily Online*, 30 October. Available at: http://english.people.com.cn/busi ness/8441900.html (accessed 17 June 2014).

Xinhua (2014) 'China Imports More Natural gas in 2013', *China Daily*, 4 February. Available at: http://usa.chinadaily.com.cn/business/2014-02/04/content_17268125. htm (accessed 17 June 2014).

Weitz, R. (2011) 'Can China, Russia Close Gas Deal?', *The Diplomat*, 22 October. Available at: http://thediplomat.com/2011/10/can-china-russia-close-ga s-deal/ (accessed 17 June 2014).

4 Low-carbon energy in China's energy security strategy

Benjamin Denjean and Cyril Cassisa[1]

Renewable energy and energy security: a desired marriage

This chapter is defined along the global definition of renewable energy (REN).[2] It is not within the scope of this chapter to look at specific technological solutions and assess whether their performance makes them truly renewable. Nor will we use the Chinese definition that distinguishes between large-scale hydro and other REN. The Chinese concept of REN includes energy with low emissions, whereby nuclear power, albeit intrinsically non-renewable, is considered a REN due to its low carbon value.

When considering renewable energy, one should bear in mind the diversity of its forms as well as the various constraints on each industrial segment. This diversity can be seen in its substitution potential, which in China is distributed differently depending on the province, as well as by the consideration of subsidies as a tool for economic development – the policy side – that is also area dependent. When the State Council adopted China's Renewable Energy Law (REL) in 2005 (China State Council 2005), which took effect on 1 January 2006, it officially recognised the role REN played in contributing to its energy security. In particular, renewable energy is presented as an important means of optimising China's energy supply mix, mitigating environmental pollution,[3] improving energy supply security and promoting social development in rural areas. This chapter will therefore describe not only the production capacity, but also the environmental security of different forms of REN and their contribution to securing national sovereignty. The high levels of air, soil and water pollution in China have now become important factors in the rising negative public opinion, a problem that the government can no longer neglect. Beyond a general concern for energy security, it is thus important to keep the issue of environmental degradation in mind, as it is currently having an increasing impact on the state's stability and security. Before introducing the historical and political framework of REN, we must remind ourselves that reforms evolve in an entwined network of conflicting interests where those of the country, indivisible and independent, take precedence over those of individuals. Yet such a dichotomy becomes blurred where the consequences of civil unrest might overcome macro concern in terms of the country's priorities.

Despite now occupying an important position in China's national political agenda, clean energy remains far more expensive and less developed than fossil fuels. The development of the green economy is still marginal within the overall Chinese economy. This switch will undoubtedly require the government to play a crucial role in restructuring economic development. With regard to the development of renewable energy, the REL specified five areas of intervention to boost its place in the national energy supply: overall targets, compulsory grid connection, classified power price, cost-sharing and the establishment of a special fund for the development of these types of energy (Wang *et al.* 2007). However, the specific measures that the REL included were insufficient in the face of the rapid development of both manufacturers and project developers, and well beyond the expectations of laws and plans (Fig. 4.1). In 2009, this led to an amended version of the REL, which was adopted to reflect the new situation of abundant supply and the increasing demand for renewable energy (China State Council 2009). Three major amendments can be identified: a more science-based approach to the formulation of renewable energy plans, the establishment of quotas for electricity generated through renewables (in order to ensure better grid connection) and the creation of subsidies (tax abatement, government pricing, concession bidding, feed-in tariff).

Although it cannot be said for all sectors – the wind and hydropower industries are already well developed in China – the REN industry is still young and keeps the ties between energy security issues and REN focused on their potential rather than on their responsibility in easing the nation's energy dependency. By 2015, the use of commercialised REN is forecast to account for 9.5 per cent of total energy consumption. However, it is important to remember that establishing a stable market demand is still one of the REN sector's principal aims. Besides hydropower, the government commonly acknowledges[4] that renewable energy still has a need for new technologies and demonstration projects. It is estimated that most of the potential for renewable energy is not being harvested due to existing challenges in the grid connection. While there is much concern is about this issue, the problem is nevertheless exacerbated by the growing excess of wind turbines and solar projects. The fact that power production from energy sources does not follow standard patterns raises growing concerns regarding the access to power of many areas, since these installations are not able to follow patterns of energy demand. This would imply a further reliance on conventional power plants that, despite their heavy environmental costs, reduce the financial burden and need for new technology in improving access to reliable electricity.

Nuclear energy, as a low-carbon footprint energy has also emerged as an inevitably strategic option towards diversifying the Chinese energy mix. In 2012, China's nuclear electricity production amounted to less than 2 per cent of total electricity production. Yet China has incontestably become the world's top market for the construction of nuclear reactors: in 2013, it had 20 nuclear power reactors in operation and 28 reactors under construction

(World Nuclear Association 2014). Further reactors are to be built which will display the world's most advanced technologies. In this context, nuclear energy is bound to play an increasingly important role in the energy mix, especially in the booming coastal areas that are far from the coalfields. Moreover, in the context of a rapid increase in energy demand, a sole reliance on renewable energy cannot help in reducing the country's coal dependency. A strong nuclear development strategy is thus essential to support the transition to cleaner energy. However, many challenges lie ahead, such as acquiring and mastering advanced nuclear technology, improving China's dubious nuclear safety culture and securing a steady supply of uranium to power facilities.

Clean energy will play an important role in China's energy transition to developing a low-carbon approach, with the aim of reducing the share of coal in the energy mix. The 12th Five-Year Plan has set very ambitious targets for all clean energies (hydro, wind, solar, biomass and nuclear) to be reached by 2020. While clean energy is an important solution to the problem of energy security and climate change, it also faces environmental and social constraints as well as technological barriers that could prevent their development. In the following sections we analyse the situation, potential risk and Chinese governmental actions for the various clean technologies.

The place of hydropower: supply or safety

China's first hydropower station was built in 1912, but the country has, over the past 50 years, become the uncontested leader in the construction and operation of large-scale hydropower (Fig. 4.2). These changes were initiated under Mao's leadership in a sustained effort to harvest nature's potential to serve industrial and economic development (Shapiro 2001).

Over time and following mass projects intended to use the full potential of every major Chinese river's full potential, the hydropower industry became an important tool for both employment and rural development. The involvement of major state-owned enterprises (SOEs) and the multiple roles played by this industry contributed to creating an efficiently developed, yet not fully transparent, industry.[5] In the case of hydropower, what is important is the emerging complex relationship between energy security and environmental security. More specifically, the dependency on water resources and its role as a water management tool have made hydropower a central part of a 'water–food–energy nexus'. [6]

Concerns arise about the possible impact of large-scale dam projects on the seismic activity in China. This is a particularly sensitive topic in provinces where it has been reported that dams have been constructed without proper Environmental Impact Assessment (EIA) or have simply lacked consideration of this aspect. The complex relationship between existing regulations and environmental law enforcement in China is already well known. Until the recent publication of the new Environmental Protection Law in April 2014,[7] this capped the maximum fine for industrials at RMB 200,000, allowing them

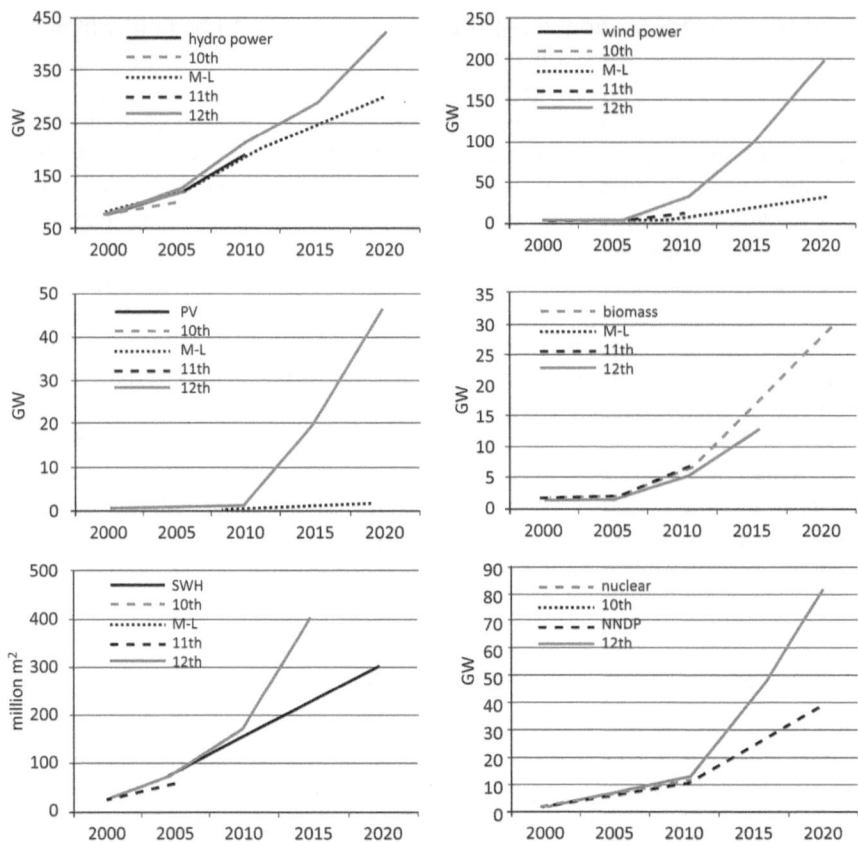

Figure 4.1 Comparison of different targets and actual capacity of clean energy technology

Source: Delman and Wang 2014.

to start construction before completing the EIA. Besides the increased risk to energy security represented by the possible damage that seismic activity could do to infrastructure – often impairing the grid connection network or the power production capacity – this hazard is included in aspects of safety. China has not forgotten the extended series of over 2,900 dam failures that followed the first wake of construction between the 1950s and the 1980s (Shapiro 2001).

While a lot of environmental NGOs have focused their criticism on large-scale hydropower plants, recent studies have shown that small-scale plants could lead to an even greater environmental impact – in particular regarding biodiversity – even for similar nominal capacity (Kibler and Tullos 2013). Nevertheless, the results are still subject to much debate, even among experts. It should be kept in mind that these studies do not take into account the various fundamental differences between small and large-scale hydropower

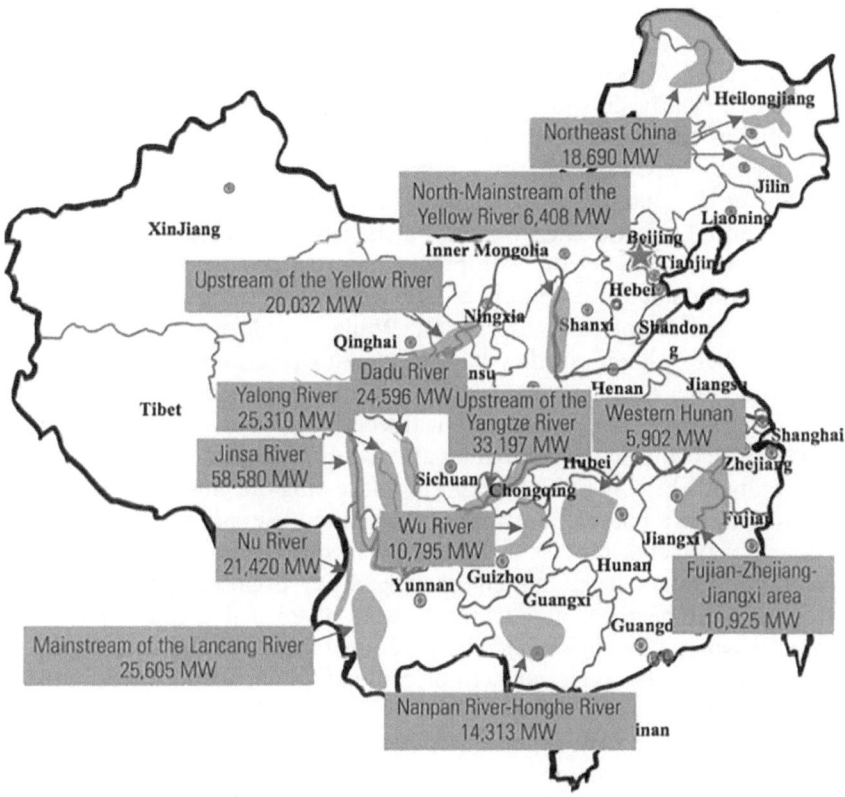

Figure 4.2 This map shows the location of China's 13 largest hydropower plants
Source: China Electricity Council (www.cec.org.cn).

plants such as flood regulation, water resource management or the associated
population displacement must not be neglected.

Multiple concerns over hydropower development

Beyond the environmental and safety problems, the power level delivered by
large hydropower production facilities renders them direct competitors of
conventional thermal power plants. While this power supply in large part
benefits from a promise of stable and predictable output, the history of
hydropower in China has proven that relying too heavily on this 'abundance of
resources' is dangerous for national energy security. The best example is
the drought that hit southwest China from late 2009 to May 2010. The
natural disaster was so severe that from September 2009 to March 2010
the total hydropower production for Yunnan province only represented half
of the previous year's production (Ministry of Environmental Protection

2010a). In order to provide electricity to the region, the government had to resort to using considerable number of coal-fired power plants, increasing the SO_2 emission by 78 per cent (Ministry of Environmental Protection 2010b). This example clearly illustrates the complex dependency on peak production energy sources to ensure a sufficient provision of energy, even in the case of existing baseline renewable energy sources.[8] According to the work carried out by Collier *et al.* (2007), such over specification of development would also predict potential adverse effects based on the resource curse.[9] Applied to this case, the increased reliance on hydropower in border provinces such as Yunnan is locally detrimental to the amount of land that is used for tourism, agriculture, aquaculture and ecological conservation. While they do not match the profitability of power production, they have met major priorities of national development and either have important developmental potential – tourism – or are already under threat – food and ecological security (Luo *et al.* 2014) in Yunnan. However, since hydropower companies are mainly state-owned and the Chinese economy already heavily relies on exports, addressing the matter of the resource curse cannot give a complete picture of the national situation. Nevertheless, it still clearly represents how both hydropower construction revenue and geopolitical importance have limited or have been opposed to other crucial developments along the major Chinese rivers. Even if it is difficult to assess the overall benefits of development, there are reports of the increase of the number of unexpected consequences of cumulative impacts – for example on biodiversity (WWF and CEFA 2013).

In recent years, the rise of multiple security issues has brought concerns about the continuing construction of dams along Chinese rivers.[10] Several reasons explain this important recourse to hydropower projects. First of all, the country has a great need for stable power production, making hydropower a provider of choice. Secondly, the Chinese hydropower industry is looking for markets both in neighbouring states and overseas. However, the possibility for China to provide electricity to ASEAN countries that is produced by dams through the ASEAN Power Grid raises several questions. One advantage of this is the possibility of selling unused electricity from low population density areas like Yunnan to neighbouring countries, as hydropower bases in Western China are power-exporting sources. It would also help in tackling the issue of increasing seasonal and temporary power shortages in those areas. However, the current driver for hydropower is still local demand, and this is the thrust for development.

New connections that are created from these exchanges could have a double effect. On a foreign relations level, Nakayama and Maekawa (2013) argue that it would improve the transition from an energy independence paradigm to energy inter-dependence; easing the political tension in the geographical area, China already has a track record of active participation and investment in important dam projects along the Mekong area. It became a key player in regional development thanks to action carried out in over 17 countries through providing unmatched funds for project financing.

According to the US Energy Information Administration, Chinese investors such as the state-owned Export Import Bank of China and the China Development Bank financed 46 per cent of all hydroelectricity capacity additions in Cambodia, Laos and Myanmar between 2006 and 2011. Hydropower know-how is thus becoming a tool for exporting development and stabilising regional foreign policy. As reported by the network of International Rivers,[11] Chinese industry and banks would be involved in the construction of some 300 dams in 66 different countries by the end of 2013.[12]

In this respect, the important cooperation between industry and the party constitutes a synergy of impressive power. Investment banks that are providing funds are also state-owned and respond directly to government policy, which increases the capacity and scope of China's international public action. Nevertheless, China's central position in the southeastern water basin does not necessarily mean that relations with its neighbours are smooth. To begin with a macro perspective, China is the upper riparian in most of the 40 major trans-boundary watercourses that it shares with 14 of its neighbours (Wouters and Chen 2013),[13] and the country still refuses to agree to any water treaty. This position is also shared by India. A good illustration of this is the case of the Myitsone hydro dam in Burma, a project that was halted in 2011 by the Burmese government because of environmental concerns.[14] Conversely, a recent report showed that Cambodia still praises the Chinese policy of non-interference in the domestic affairs of recipient countries (Grimsditch 2012). This policy, known as the 'Dam policy', has proved to be a significant and complex issue for China and its neighbouring states. Its potential became so important that the USA started weighing into the issue and support countries of the lower Mekong in their claim for increased water flow, delaying construction of mainstream dams (Senate of the United States 2011).

Besides the Mekong River itself, China's territory encompasses three more of the greatest South-Asian trans-boundary river sources, notably the Yangtze, the Nu River and the Brahmaputra, strengthening its position of 'upstream superpower' (Nickum 2008). It is only very recently that India, sharing similar development needs as China, agreed to a common development plan. However, growing tensions on the subject of the acceleration of India's infrastructure completion emerged in September and October 2013, the South Asia country racing to claim the right of 'first use' in international law.[15] If India had continued to adopt this stance, it would have firmly stepped out of a tradition of run-on-the-river[16] projects and raised the issues of town submergence and population displacement. Eventually signed in late November 2013, the agreement to ease local tensions brought cooperation on data sharing – both urgent and necessary – but did not improve any of the strategic issues. This dispute over the control of the Brahmaputra River holds a very strategic value, as controlling its flow would not only give the upper hand on Bangladeshi and Indian farming, but it might also incur potential floods over downstream populations. This fact clearly highlights the international policy component and the national security issues in hydropower management. To follow the example of India and to better understand China's position, it is necessary to take a closer look at China's claims, in other words the 'Principle of Absolute Territorial Sovereignty' (meaning that the upstream state's interest always prevails). The corresponding mindset was very well depicted by Ambassador Gao Feng's declaration at the UN General Assembly in May 1997, where he explained why China opposed the UN convention on shared water states: 'The text did not reflect the principle of territorial sovereignty of a watercourse state. Such a state had indisputable sovereignty over a watercourse which flowed through its territory'. As such, it maintains a clear power-dependency relationship with most of its neighbours.

Tensions with neighbouring countries are especially high when it comes to the broader context of climate change, even if the interpretations given to this issue may be different. For instance, in 2011, Zhang Boting, the Vice-Secretary General of the China Society for Hydropower Engineering believed that droughts in southwest China only confirm the need for more hydropower facilities in the country (Cho 2011), an opinion not shared by all experts. For Chen Guojie, a researcher at the Sichuan Mountain Hazards and Environment in Chengdu, forecasts of long-term energy provision through dams are based on current natural conditions, anticipating this availability for the next 100 years. However, conditions change quickly and neglecting the impacts of climate change would mean overlooking their harm on the environment and the possibility that they 'might very well prove dysfunctional in the not-so-distant future' (ibid.).

Wind: an unstoppable advance

Over the past 20 years, China, incontestably the most important market for wind power, has witnessed the wholesale creation of an industrial sector, from its early domestic beginning through to its expansion into international markets, and finally ending up being the main pillar of Chinese renewable energy and low carbon development policies. Before entering into detail about present wind power specifics and its implications in Chinese energy security, it is important to understand the extent of the unprecedented growth of this energy industry.[17] Such a description is necessary in order to understand the interdependency between the access to foreign markets, local subsidies and energy independence.

Unprecedented – and uncontrolled – growth

China was one of the very first nations to design windmills used to harvest the power of wind (Adam 2006). However, the country had to wait until 1994 to employ wind energy as a form of power generation. It then entered the so-called 'Wind Power Generation Industry Development Phase' that lasted until 2003. This move resulted in a steady growth, until 2005 when installation started to increase at a dramatic pace. From 2005 to 2012, China expanded its wind energy capacity by almost 50 times. Moreover, windfall profits from feed-in tariffs and off-grid wind farms that were too standard were hindering wind power credibility and overall economic performance.[18] In order to take regional disparities into account, the National Development and Reform Commission (NDRC) released the 'Circular on Improving Wind Power On-Grid Tariff Policy' (National Development and Reform Commission 2009) in 2009. The circular provided each province with one out of four feed-in tariff classes as shown in Fig. 4.3, marking the start of a period that is considered 'reasonable growth', with a rate falling around 50 per cent. This gave wind power a longer-lasting rational place in the Chinese energy mix.

As a result of these policies, wind capacity was already accounting for 63 gigawatts by the end of 2011,[19] which brought more wind capacity to be installed in five years (2007–2011) than either the USA or Germany, yet both of their developments took over 30 years.[20] This number also widely surpassed the national development plans. Original objectives accounted for 50 gigawatts in 2015 and 100 gigawatts in 2020. New political commitments from the 12th Five-Year Plan then doubled the objectives for developing the sector by 2020 (Fig. 4.1).

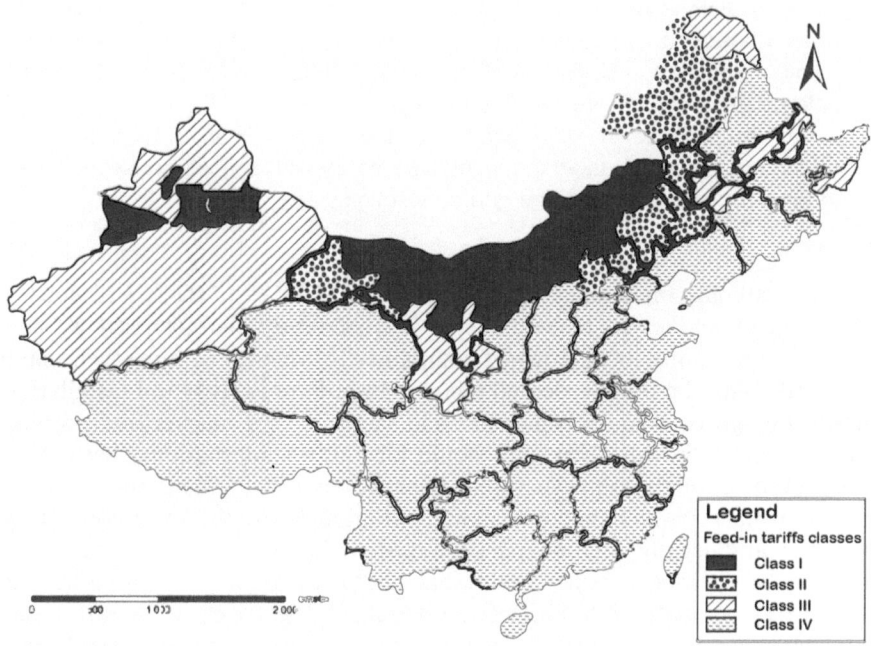

Figure 4.3 National division of feed-in-tariffs per province
Note: Even with those differences in feed-in tariff, Inner Mongolia still remains the most important host of wind power projects.
Source: Authors' own drawing.

Wind development in China thus announced an exceptional situation, where industry was over five years in advance on the objectives set by national plans. Such overcapacity is also explained by the so-called 'resource for industry' principle which forces industries to purchase high levels of locally produced equipment[21] (not only at national, but also provincial or even community levels). The installation of local factories was expected to have a much greater impact on the GDP growth of the administrative units, which is once again the main lever towards the promotion of local government officials. This eventually led to the creation of a large number of partially vacant factories. In 2012, the Chinese market ultimately registered a drop in sales, which resulted in a global shrinkage of installation and lead to the disappearance of Chinese manufacturers from the top five turbine manufacturers in the world.

Box 4.1 The flaw in the words of law

Unfit for the development of concentrated wind farms in China, several projects were built in agreement with the '49.5 megawatts wind farms' effect. According to Lu Hong, an expert in renewable energy at the Beijing office of The Energy Foundation, many wind power projects in China are built with a capacity slightly below 50 megawatts – the power threshold at which a project needs to be granted permission not only from local authorities but also from officials in Beijing. In other words, their feed-in tariffs are no longer solely determined by provincial authorities.

This similarity to the 12 megawatts wind farms in France illustrates the strong impact of threshold in the industry policy with serious consequences in terms of both planning and grid connection.

International consultancy agencies such as MAKE or BTM still consider the Chinese market to be weak,[22] as it is currently characterised by fierce competition among locally owned companies to which it is, moreover, mainly reserved. With the unprecedented increase in both objectives and installations, major foreign companies began to build factories in a market that promised more domestic buyers than they would have at an international level. However, even through joint ventures, these foreign manufacturers could not secure a market share that was significant enough. As the local content was the first and legally binding part allowed to be sold on the market, most companies did not consider additional unofficial binding or exclusive partnership agreements with project developers.[23] Moreover, while most local private companies are willing to form partnerships, most successful actors are still SOEs, leading them to choose local manufacturers first.[24]

Nonetheless, China was also expected to have installed less wind power capacity in 2013 than in the previous year. This market shrinkage not only affected foreign manufacturers, but also the industry as a whole. In response to this situation and to the broader issue of turbine oversupply, the government implemented the objective of developing 'Wind Bases', gigantic projects that are to secure policy support in order to dampen pressure on manufacturers (see Fig. 4.4). This measure is to be implemented in relation with the 2013 announcement of various 1 gigawatt solar projects that answer to a similar need. This proves that wind industry in China is still a market that is driven more by supply than by demand.

Improving the quality of the equipment while driving benefits from low prices

Chinese manufacturers have historically suffered from their poor reputation compared to Western and even Indian competitors. Even the Chinese State Electricity Regulations Commission identified construction issues and accidents as a significant problem within the local market, whereas the vast majority of players compete for the reliability of their machines (Li *et al.* 2012). In 2011, this concern led to a shift in building up the industry's reputation for quality.

Due to the widespread use of calls for tender procedures, Chinese manufacturers have also recently drawn benefits from their low prices. The fierce competition among the various companies has dramatically brought turbine prices down, from RMB 6,700 per kilowatt in 2009 to RMB 3,500 per kilowatt in 2011. Since the local market could not absorb the entire turbine production, China started to look into foreign opportunities. As was previously mentioned, this issue mainly concerned local players, and it is principally

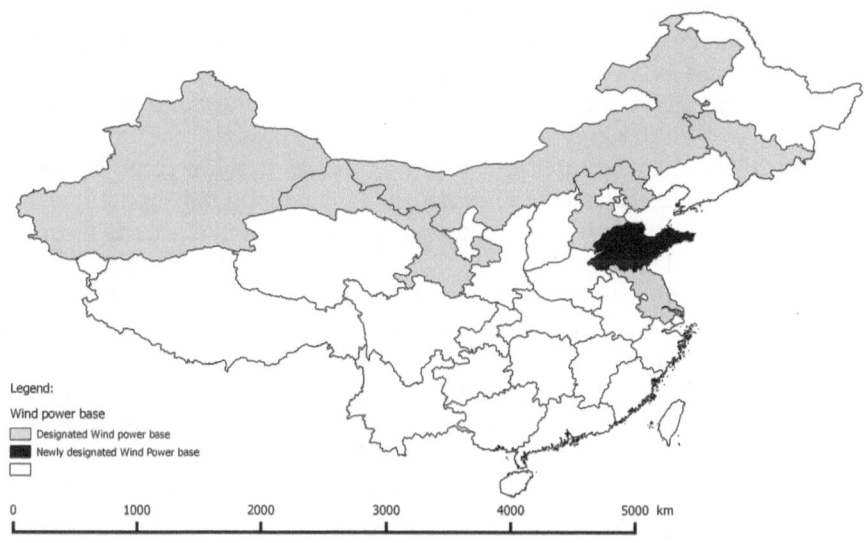

Figure 4.4 Localisation of the Chinese wind power bases
Source: Authors' own drawing.

operated by state-owned companies or the central government. As a result, the tightly planned and subsidised wind industry cannot be seen as a long-term strategy for exports. China concentrated its effort on two different axes in particular, namely the careful acquisition of the know-how of top players through joint ventures and foreign investments in power generation. The first axis helped to create the competitive wind industry as we know it today, clearly exemplified by the acquisition of the Danish Vestas' tower manufacturing plant in 2012 by Titan Wind Energy (the Chinese wind towers manufacturer). Titan Wind Energy was Vestas' supplier for a long time, but thanks to this transaction they have now become co-owners of Vestas' manufacturing facility in Denmark.[25]

A second strategy is using foreign investment to expand to international markets. This is clearly separated from the former strategy (Tan *et al.* 2013). Sixty-three per cent of the investments are indeed direct investments in power generation. Such a strategy holds multiple aspects: first of all, this investment policy is highly supported (when not imposed) by the State. Secondly, it enables consortiums and joint ventures with strong ties to the Chinese industry to be established, allowing a subsequent (if not directly through concomitant calls for tender) easier market penetration for national production. Finally, and less centred on national energy policy, it puts countries that are on the receiving end in debt to China for their energy production, creating policy leverage. The example of wind power places renewable energy as an element in the Chinese soft power strategy, ensuring international security by interdependency (even if not symmetrical). It is noteworthy that the Chinese appearance in international markets does not focus here on innovation or expertise like Western players do. While most of them have a strong Intellectual Private Property (IPP) and 'knowhow preservation' policy, China plays what could be seen as a counterproductive game by selling its newly acquired technology to local companies in emerging markets. As an example, Isivinguvungu WindEnergy Converter's (I-Wec) became the first South African turbine manufacturer by acquiring blade moulds from China in October 2011. Nevertheless, the purpose was to create a steady and important foreign pipeline that aims to keep economies of scale, and thus affordable prices at the local level. The solar situation is strongly linked with the fact that the employment aspect brings an important component towards preventing social unrest and thus public security, which is at the heart of China's development priorities.

Nevertheless, many constraints still exist, limiting the possibility of investments for Chinese groups abroad. In November 2013, King & Wood Mallesons, a global law firm with its headquarters in Asia, reported that the Chinese-owned Ralls Corporation in the USA was denied access to buying a wind farm located near a navy area in Oregon, USA, and national safety was cited as the reason for the rejection.

As analysed by Tan *et al.* (2013), even if the investments are now booming, the majority of foreign investments in both wind and solar industries have been in developed countries (Tan *et al.* 2013), as illustrated in the case of Wind Industry in Fig. 4.5. More than just an ability to absorb overcapacity, the strategic choice of Chinese players to absorb the political risk in emerging countries is a way to gain international visibility for the quality of their production markets that are yet to be fully developed.

Grid capacity limits to the expansion of wind energy

A noteworthy limitation to the expansion of wind energy in China derives from the specifics of wind development, mainly constituted by large-scale, centralised plants and both long-distance and high-voltage transportation (Zhang *et al.* 2010). When it came to grid integration, this resulted in problems arising sooner than in foreign countries. Countries with a more spatially distributed production capacity indeed benefit from complementarities of wind regimes and dispatching in diverse network nodes. This allows the grid to withstand higher content of wind energy without requiring any investment in upgrading. In China, developing a vast network of connections and implementing a substantial level of innovative technology in the deployment of the grid both became necessary. However, questions arise as to the additional difficulty of reforming the grid system, which was designed for coal-dominated production. Passed in 2011, the new regulations regarding wind power aimed at tackling the issue of grid availability by requiring all new projects to be reported and approved by the central government before having either feed-in tariff or grid access. Such a move enabled a more thorough global supervision and planning of the wind farm installations, by matching the grid capacities with the production capacities. The difference

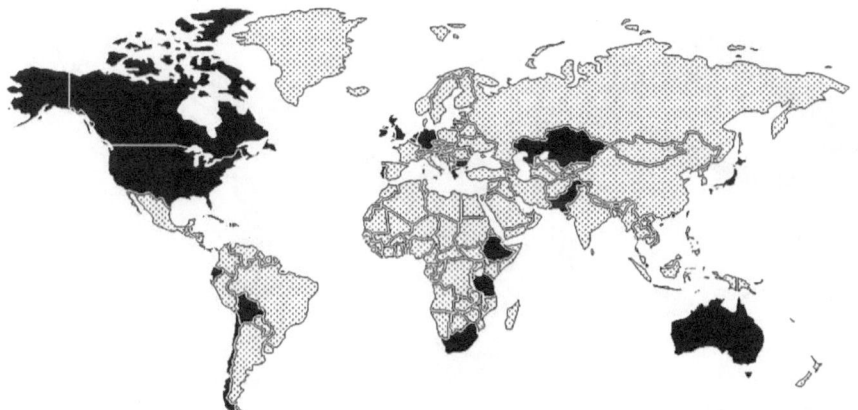

Figure 4.5 Overview of China's overseas investment destinations in the wind industry
Note: The black countries represent the use of Chinese investments in the wind industry.
Source: Data from Tan *et al.* 2013.

between grid capacity and power production led to an average curtailment of 12 per cent worth RMB 6.6 billion in 2011. Tackling this issue is therefore necessary in order to achieve the final production objective of the Five-Year Plan, entailing a global improvement of the capacity factor by 15 per cent. Since 2010, difficulties in coal fire business, combined with the aforementioned curtailment and accident issues, significantly complicated the access to loans for local projects. As such, the actual integration of large-scale wind power into the Chinese electrical grid could be seen as a threat to Chinese energy security.[26]

It is also worth mentioning that Class 3 turbines[27] have not recently been available on the Chinese market. The industry thus faces a situation with a maturing industry that will still be subject to technological changes with a major impact on planning. However, this again raises the question of international cooperation and technology transfer, as since 2013, only foreign manufacturers master the design and manufacturing of large blades.[28]

During the start-up phase, wind power development was mainly concentrated in the 'Three Northern'[29] regions. Yet Fig. 4.6 clearly shows that competitive wind power from Class 3 turbines would imply unlocking new sources of energy for densely populated areas. The close proximity of both industry and high demand is one of the major drives for the second phase of

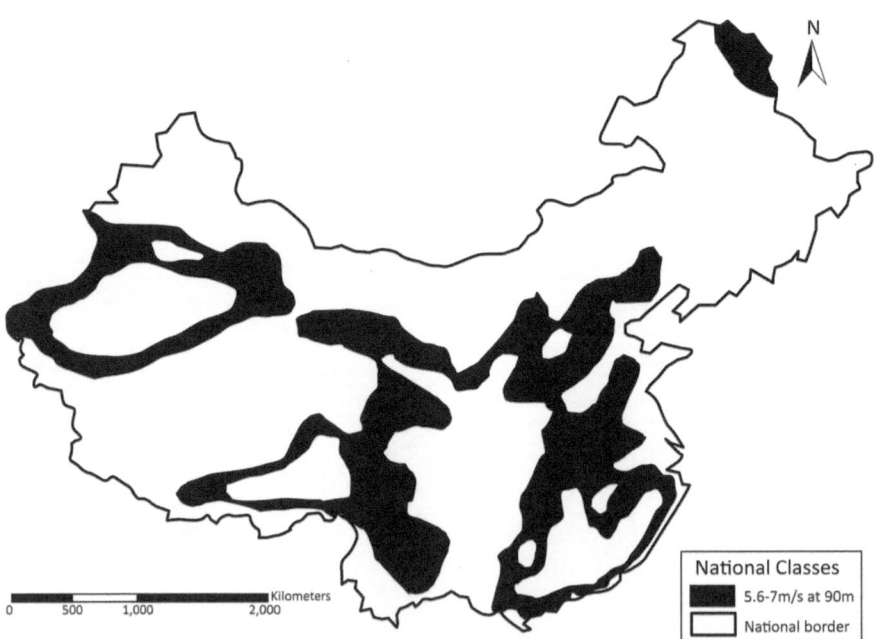

Figure 4.6 Areas fitted for the installation of Class 3 turbines
Source: Authors' own drawing.

wind development as identified by Liao *et al.* (2010). It is precisely here that wind technology's potential for energy security unfolds. From this stage of development, it stops being merely an additional energy source or employment provider and starts providing new solutions in areas that are deprived of natural resources. This echoes the previous limitations, which appeared in the capacity of wind energy, to address the issue of equity in resource distribution. Before these technologies, wind farms were concentrated in areas sparsely populated and rich in coal. The intended 'Wind Bases' indeed remain in areas sparsely populated that are often rich in natural resources (with the exception of Hebei).

Solar energy: great potential with technological constraints

China is one of the richest countries in solar radiation. The Europe–China Centre for Clean Energy (EC2)[30] estimates an annual radiation of 5.10^{22} joules with an annual surface absorption heat equivalent to about 1,700 billion tons of coal. Assuming that 20 per cent of roof area and 2 per cent of the Gobi desert and other remote areas are installed with solar power facilities, China's exploitable capacity could reach 2,200 gigawatt. Against this backdrop, the potential of solar energy will probably play an important role in Chinese energy security in the near future. However, even with a strong growth over the last decade, the solar energy sector is not yet developed enough and its scale remains almost negligible in the Chinese energy mix.

A strong photovoltaic industry to save

Pulled by the European market, China's photovoltaic cell manufacturing industry rapidly developed after 2004. During the 11th Five-Year Plan, the photovoltaic (PV) industry became one of few Chinese industries to compete globally and be expected to reach an advanced level internationally. Since 2007, China has been the largest manufacturer of PV cells in the world, experiencing an annual growth rate of over 100 per cent. In 2010, production of solar cells was around 10 gigawatt, accounting for 50 per cent of the global production. The great difference between the 12th Five-Year Plan targets and former targets of 2005 reflected both technological and market breakthroughs (Fig. 4.1). Energy conservation and a reduction in emissions in the PV industry achieved remarkable results, with a steep drop in costs, going from US\$4.5 per watt in 2000 to US\$1.7 per watt in 2010 (National Development and Reform Commission 2012).

This strong growth of production capacity could be explained by Chinese 'push nature' policies (Wang 2013), passed in 2006 and providing fiscal facilities and tax abatements for PV R&D,[31] and in 2008 providing income tax abatement for PV production firms during the first six operational years.[32] During this time, other countries in the EU primarily focused on stimulating the purchase of PV panels, contributing also to an increase in Chinese PV production. However, PV prices still remain higher than other energies, which limits its global expansion. In 2012, the global production capacity of solar modules was about 70 gigawatts, while the global demand did not in fact exceed 36 gigawatts (Laurent 2013).

Adding to the overcapacity, the Chinese PV industry mainly relies on overseas markets and, as a result, the development of the industry has largely been affected by the global financial crisis and changes in overseas markets. More than 90 per cent of Chinese solar cell products are exported, while the European market accounts for about 70 per cent of the global market. However, the absolute dependence on the European and US markets resulted in a fatal blow to Chinese solar PV manufacturers. Anti-dumping duties were imposed by the USA in 2012 and by the European Commission (EC) in 2013, seriously harming Chinese PV cell and module producers (Delman and Wang 2014). As a result, this overcapacity of Chinese PV production, coupled with international trade disputes, destroyed industry profits. In 2011, one-third of Chinese PV manufacturers were shut down and nine Chinese solar manufacturers suffered huge losses (the biggest company, Sun-tech Power, lost close to US$1 billion).

To keep the industry alive and to increase the role of PV in Chinese energy security, focus was directed to the domestic market. Recently, the development of PVs has been significantly encouraged by the government, which is likely to adjust the solar PV installation target to up to 35 gigawatts in 2015. Moreover, the government simultaneously introduced 'pull nature' policies (Wang 2013). In 2009, for instance, the Power Building Subsidies Fund and the Golden Sun project were introduced in order to finance and subsidise PV power building and electricity projects. In 2011, interim measures for the 'Administration of Financial Subsidies for Application of Solar Building-Integrated PV' and a notice on implementing the 'Golden Sun Project' were released to scale the PV power projects up in China. It is thanks to these measures that PV producers and PV electricity power investors are now able to receive subsidies from the central government. In 2011 and 2013, China established detailed rules for feed-in tariffs (RMB 1 per kilowatt-hour except for in Tibet, where it represents 1.15). China also insists on combining grid-connected power generation and off-grid application. In 2011, the NDRC released a standard price regulation for PV power grid connections, whereas since March 2012, subsidies have been provided to PV electricity power grid connections through financing the project of grid connection. Finally, the State Grid has released a policy to provide several facilities and free or low cost services for PV electricity grid connection.

With regard to PV technology, China still lags behind international advanced levels of polysilicon key technology: high-end equipment used for manufacturing crystalline silicon cells still needs to be imported, and Chinese enterprises notably lag behind in thin-film cell processes and equipment (National Development and Reform Commission 2012). The Chinese government is therefore encouraging leading PV enterprises to expand through technological progress with the aim of narrowing the gap with international advanced production technology, focusing on silicon-based PV, grid and energy storage technologies. China is also promoting its 'going out' strategy and participation in international competition. China's PV industry abundantly utilises both domestic and foreign capital and human resources. By the end of the 11th Five-Year Plan period, dozens of enterprises were listed nationally or abroad, and their products are now sold globally. Domestic PV enterprises are predominantly privately owned, differentiating between the solar PV sector and the wind sector, giving leading enterprises the ability to grow stronger. After completing their vertical integration, leading enterprises accelerated overseas mergers and acquisitions and developed into international enterprises that assured them foreign technology acquisitions and stability in the international market.

From the perspective of the strategic path of China's social and economic development, promoting the solar PV industry is essential for guaranteeing energy supply and establishing a low-carbon society, for promoting economic restructuring and fostering strategic emerging industries. With all these new policies to promote PV cost abatement and domestic market development, there is no doubt that PV power will play a significant role in the Chinese electricity mix as much as or even more than wind energy. By 2020, the central government is expecting the PV power generation to become economically competitive, costing around RMB 0.6 per kilowatt-hour.

Solar thermal, the world's largest capacity in China but still underdeveloped

In China, solar thermal technologies have not yet reached the same stage as PV technologies. Thermal energy still mainly relies on fossil fuels. Solar energy for thermal use, which includes solar heaters, solar thermal power generation among others (e.g. seawater desalination, solar house, cookers), can significantly reduce the share of fossil fuel.

China currently represents a large majority of the world's solar heating market and installed capacity. The nation has now reached an estimated 310 million square meters of collector area. This represents 275 gigawatts of thermal capacity, more than double the capacity of the rest of the world's installations (Epp 2014). There are three principal categories of solar water heater applications in China: the retail business of systems for single-family homes in rural areas, the project business in cooperation with property developers in urban areas, and the medium-temperature applications in the industrial sector. The first two categories represent the majority of the market; with the support of the Chinese government, the third might grow at a rapid pace in the future. The inventory volume of China's solar water heater industry is expected to grow to 400 million m^2 by 2015 and to 800 million square meters by 2010 (Qianzhan Co., Ltd. 2014).

However, the central government has only recently started to promote solar thermal power generation. Given the high annual radiation in China, the solar thermal power industry should grow quickly in the near future, leading to additional policy support on behalf of the government.[33] The first tower-type solar–thermal power station in Asia was built in the northwestern area of Beijing. Compared to traditional thermal power stations, this solar-thermal station should reduce annual emissions of CO_2 and SO_2 by 2336 tons and 17.5 tons, respectively. The Government of China has announced a target of 1 gigawatt of concentrated solar thermal power (CSP) capacity by 2015 and of 3 gigawatts by 2020.

China dominates the global solar thermal market by monopolising 64–69 per cent of the existing solar heating and cooling capacity. In 2010, China's paper-making, food, tobacco, wood, chemical, pharmaceutical, textile and plastics industries consumed 450 million tons of coal, used mainly for heating and drying. If 10 per cent of the energy needed by these industries emanates from the sun, then it would be indispensable to have 300–400 million square meters of heat collectors (Song 2012). This shows the need for improved efficiency, as well as promotes market development. Future priorities include testing and certification systems, better quality control, better integration into architectural practice and the construction industry, product standardisation, consumer education and awareness, new financing mechanisms, and favourable tax and tariff policies. The turnover in the solar thermal industry was approximately US$17 billion both in 2012 and in 2013, which is five times the size of Europe's solar thermal industry.

Biomass: from rural necessity to urban cleaning

China has historically been attached to its vast use of biomass for energy provision, and it achieved an unprecedented use of digesters for its rural population. However, new social standards, together with the craving for urban living standards, encouraged the distorted image of digesters in rural areas. This tendency makes it difficult to maintain or promote this solution in poverty alleviation projects, as highlighted by Planet Finance,[34] a leader in providing microfinance services in China.

Agricultural biomass: limitations to its full potential exploitation

Considering both agricultural and animal waste (poultry and livestock manure), together they accounted for nearly 75 per cent of China's biomass resources in 2010 (Zhao *et al.* 2012). However, even though both resources and technology are available, policies regarding unsuitable energy pricing make it difficult to have a stable market development for this resource. If we look at the theory of an industry's development life cycle, bioenergy in China is at the transitional stage between start-up and growth (Zhang *et al.* 2009). Only when limitations, such as a lack of coordination, issues in feedstock supply and long-term policies and regulations are solved, can this industry continue its expansion and then actually provide a valuable source for China's energy provision.

To date, the sector is driven by five major state-owned power generation groups, whereas private investors still only represent a minority of investment (Zhao *et. al.* 2012). While this concentration shows the political dedication to the development of this form of energy, it also highlights that it is not yet a sustainable energy that is marketable.

An additional limitation resides in the spatial distribution of industrial exploitation. Leading projects are currently being developed in regions such as Inner Mongolia, where population density is very low, creating a gap between individual consumption, which is concentrated around developed and highly populated eastern provinces, and energy production, which is concentrated more in the western areas. Similarly to natural and unconventional gas industries, gas produced through biomass also faces problems of a network that is both underdeveloped and unreliable. Even if gasification markets are still a promising opportunity in the future at an industrial scale, they remain marginal in current power generation as opposed to direct combustion of straw or waste incineration.[35] It is however noteworthy that in order to reduce dependence on coal, new projects are developing in China that combine the use of biomass and coal.

Last, but not least, access to raw materials is also an issue in a country where nearly half of the population still lives in rural areas with very high efficiency in biomass use. A full exploration of biomass potential would imply a competition between modern renewable energy sources and previous efficient use of local resources.

Municipal solid waste incineration: can enforcement secure environmental benefits?

The subject of urban solid waste burning is increasingly important in China's renewable energy panorama, both regarding concerns over environmental safety and the availability of raw materials, as well as its potential in energy production. Vis-à-vis waste treatment, China only used landfills as a means to treating its wastes until the year 2000, creating gigantic areas to absorb the growing production of large cities. Later, when faced with the lack of capacity for safe disposal and the close proximity to highly populated urban areas, an important health safety aspect started to emerge, largely surpassing the benefits of the energy provision component for waste burning. Nevertheless, the 12th Five-Year Plan prioritised the construction of new processing facilities, making the proportion of waste burning jump from 7.5 per cent in 2005 to nearly 20 per cent in 2009, and this growth has not ceased.

Zhang Yi, the Head of the Shanghai Environmental Sanitation Engineering Design Institute, reports that industrial projects with the objective of building municipal solid waste plants have already started finding social opposition. However, these movements mainly occur in China's wealthiest areas, where environmental awareness and expectation have dramatically increased (Yu 2012). The distinctions in the opposition population category are not solely explained by social or educational differences, nor by various perceptions of towards each kind of energy – besides their environmental impact (Ansolabehere and Konisky 2009). Indeed, it also comes from the very difference in how much land is required for each type of energy. While hydropower essentially faces population displacement in rural and underdeveloped areas, municipal solid waste plants intrinsically aim at highly dense and capital-intensive urban areas where the introduction of industrial elements happens within closed spaces. The construction of such plants may also have potentially adverse effects on the land value of most developed urban areas, clearly facing the 'Not in my back yard' (NIMBY) effect (Zhou and Chen 2012).[36] This is once again developed in urban areas where speculations about real-estate price variations have important consequences on urban populations with high mobility. It is interesting to note that while its consequences are widely unpredictable, the recent Environmental Protection Law of the People's Republic of China takes a great leap forward in public participation and information disclosure, widely expanding the possibility for populations to both have access to industrial data and to start action in courts.

Box 4.2 Xie Yong: the hard path to transparency

Xie Yong lives close to a trash incineration plant in Nantong, Jiangsu province. He was convinced that his son's severe health condition was due to improper pollution control at the power plant. After a first negative judgment at the county-level trial in 2011, he decided to file a request for emissions data from the local environmental bureau. This request was denied, as was his later one to the provincial Ministry of Environmental Protection.

This brief description intends to illustrate how a legal framework does not necessarily imply law enforcement, and that environmental safety, clearly opposed to financial efficiency, needs transparency to be ensured.

Even though burning waste can appear to be a solution that addresses both environmental and energy security concerns, regulations are not enforced often enough to favour such an evolution. A complete set of standards already exists for the planning and lifetime of all power plants (Zhou and Chen 2012). However, the reason why a part of the population is concerned is the focus on control and monitoring of flue gas released from burning activities (Hong *et al.* 2010). Incinerators are subject to emissions regulations that are much less stringent than regular power plants.[37] Moreover, certain industry insiders have reported the use of coal as the raw material to boost the output of the power plants. As this can be a standard procedure in most places in order to ensure stable production, levels – when controlled – can often exceed regulations.[38] Looked at from a more political perspective, the sector is still government-driven. In 2004, the 'Municipal Utility Franchise' was established, pursuing the creation of a market and the potential to finance the waste treatment industry. However, the management of waste still suffers from a crucial lack of transparency. Interviews carried out in 2013 with local researchers have reported very limited access to installation and monitoring. As with most integrated management, an efficient system would need clear and reliable data management in order to detect any deviation from regular use and to monitor performance.

The fifth 'renewable energy': nuclear power

Modernising China's nuclear fleet

China's first steps with nuclear energy took off with its nuclear weapons programme, developed in cooperation with the USSR in the mid-1950s. This was followed by a period of inconsistent development of civil nuclear power in the second half of the twentieth century, during which different types of nuclear plants were sporadically built across the country. At the dawn of the new millennium, Beijing decided to develop its nuclear sector at a faster pace. In 2003, the Communist Party's National Congress revised the earlier guiding principle of 'appropriate development' to 'vigorous promotion' of nuclear power. Three years later, the State Council approved the goal to increase China's nuclear electricity production capacity fourfold – to about 40 gigawatts – by 2020 (National Development and Reform Commission 2007). This impetus to develop nuclear power would have in part been based on China's gradual mastering of Generation II (Gen-II) technology reactors.[39] Indeed, between 2005 and 2010, 20 or so Gen-II reactors modelled on indigenous Chinese designs entered the construction phase. Most importantly, China decided to embark on the wide-scale adoption of the world's most advanced reactors, dubbed Generation III (Gen-III).

A combination of domestic and international factors explains China's strategy in pursuing this technology. In the early 2000s, China was experiencing rapid economic growth along with increasingly high rates of electricity consumption: in 2003, as a result of the energy supply gap, 18 out of China's 31 provinces experienced electricity blackouts (Yeh and Lewis 2004). Within this context, Hu Jintao and Wen Jiabao, the new technocrat leaders at the time, recognised the nuclear sector's untapped potential. On the international scene, China was working hard to prop up its international political standing, in particular in the field of energy. China signed the Kyoto Protocol and put its civilian nuclear programs under the supervision of the International Atomic Energy Agency (IAEA). In this light, the competitiveness of its nuclear sector became a serious matter. As Gen-III technology became available, the leadership decided to use it as a foundation on which to base the modernisation and standardisation of its nuclear sector. In relation to their predecessors, Generation III reactors offer significant advantages: their design is simpler, they use fuel more efficiently and they reduce nuclear waste. Most importantly, this technology provides the latest standards in safety.[40]

In September 2004, China put out a call for tenders for the construction of two Gen-III reactors. Bids were received from the American company Westinghouse (AP1000), the French company Areva (EPR) and the Russian company Atomtroyexport (VVER-1000). This pursuit of particular interests by different players, as well as the competition between key international vendors conscious of the Chinese market's potential, have prevented a standardised nuclear fleet to materialise (Ramana and Saikawa 2011). This problem entails significant disadvantages, such as higher construction and operating costs and pricier and longer safety checks. It is therefore likely that rationalisation and standardisation will be emphasised in the future, and since the Chinese leadership settled on Westinghouse's AP1000 to modernise its nuclear fleet, it is unlikely that there will be another bidding process. Nevertheless, the repetition of isolated deals along the lines of the French import (bilateral packages negotiated between top political leaders) or the Russian import (a direct externality of China's strategic partnership with Russia) should not be excluded.

Reverse engineering and exporting nuclear reactors

By embarking upon Gen-III, China resolved a significant problem: property rights. Indeed, the property rights of China's most common reactor, the Gen-II CPR-1000, had been retained by Areva, which limited export potentials. In contrast, the deal with Westinghouse involves an important agreement on technology transfer. The official model for future home-built reactor designs, the CAP1400, is based on the AP1000 but will display Chinese intellectual property rights.

Technology transfers have been a controversial issue. It is no secret that China is aiming to build its own nuclear reactors, and for this purpose has been aggressively seeking to acquire nuclear expertise and knowhow, which in turn has engendered a stark dilemma for foreign firms. Offering too little will ensure being cut out of the market without taking advantage of the present demands for technology. However, accepting Chinese requests entails the potential risk that Chinese firms will be able to reverse engineer the construction of nuclear plants and wholly indigenise production, which would not only shut doors to the Chinese market, but would also jeopardise worldwide market shares by creating fierce competitors.[41] Actually, the view that Chinese firms will remain dependent on their Western counterparts is not unfounded. If Chinese actors have accumulated experience in plant and reactor construction and component manufacturing, they still rely on foreign firms for special equipment manufacturing, plant monitoring software or simulation systems.[42] For researchers at the China Economic Review, achieving complete localisation is 'not only a long process in terms of technical capabilities', but it will also require a 'rethink of Chinese enterprises' capacity for innovation and internal culture' (China Economic Review 2013). The sophistication of nuclear technology is the ultimate test to confirm whether China can successfully copy the West.

The indigenisation of a Chinese Gen-III reactor would not only shore up China's energetic independence, but also pave the way for a revolution in the supply of nuclear energy. Beijing is eager to become a recognised nuclear exporter, and it is likely that this aspiration has been a key drive in the fast-paced development of the sector. Domestic demand enables China's state-owned nuclear firms to qualify an increasingly large number of local suppliers, with the ambitious aim to sell cheap, homemade reactor designs and technologies at a global scale. The economic benefits could be tremendous. More significantly, leading the pack in the export of nuclear technology would become a valuable leverage in international politics. One could look to the Persian Gulf, where nascent cooperation with Saudi Arabia over nuclear energy could be part of a broader Chinese strategy to secure oil resources. It thus remains to be seen whether China can establish itself as a responsible nuclear vendor. A challenge will be to resist those short-term geopolitical and economic temptations and restrain from exporting outdated nuclear reactors to developing countries that are not yet ready to welcome nuclear technology (Tu 2012).

The impact of Fukushima on China's nuclear expansion

Before it can export its reactor, China must overcome internal issues related to its safety culture. In the late 2000s, the Chinese leadership decided to pursue the development of nuclear power at a more aggressive pace, vastly increasing its target of nuclear energy production for 2020. This ambitious plan of nuclear expansion impressed audiences, but also raised concerns. China's nuclear reactors have operated without any reported serious safety incidents since the first reactor was connected to the grid in 1991, yet echoes of China's underdeveloped security culture have surfaced. Shortages of fuel and equipment, bottlenecks of human resources and risks due to outdated technology loomed on the horizon, increasing the probability of a nuclear accident, which would have been fatal for China's nuclear sector and a serious blow to China's broader energy security policy.

The accident at Fukushima in March 2011 made the Chinese leadership strikingly aware of the potential risks of nuclear power. Following the disaster, the State Council halted projects due to start construction and suspended the approval of new projects until all nuclear sites had been inspected. The year-long security review proved mostly satisfactory, but still underlined important shortcomings related to nuclear safety. Certain facilities were found to not be meeting safety standards for flood and earthquake resilience, while certain others did not have sufficient measures to deal with major accidents such as a tsunami. The report therefore required the phase-in of safety improvements in various areas by 2015, ranging from preventing and mitigating major accidents to implementing probabilistic safety analysis for external events (Zhou 2013). At present, the green light for new construction projects would only be given to Generation III reactors, while the construction of inland plants would be put on hold until 2015. In addition, these inland facilities are at risk of seismic activity. Most importantly, their plants require a reliable supply of water for reactor cooling that China's low freshwater resources do not always provide (Wang, Y. 2013).

Nuclear construction hence gradually commenced, notably with the successful application of the first dome of the AP1000 reactor in Sanmen in 2013. Nevertheless, crucial security challenges remain. A paramount concern stems from China's generally poor construction quality. Unsatisfactory planning, shoddy quality control and unqualified construction workers all contribute to a lower cost, but also to lower safety standards. An analogous point is that while plant staffs all comply with regulations and laws, there is no culture of proactively protecting and improving the system (Zhou *et al.* 2011). Lastly, safety oversight mechanisms remain one of the weakest links of the Chinese nuclear industry. The National Nuclear Safety Administration (NNSA) also lacks resources: its R&D branch is insufficiently funded to set its own safety standards, and its under-staffed inspection teams will soon be unable to cope with China's nuclear expansion (Zhou *et al.* 2011).

In addition to improving its security culture, the Chinese nuclear sector will need to communicate more effectively with the Chinese population. Indeed, the growing number of protests related to environmental issues has proved to be a challenge for Beijing. In 2013, a plan to build a US$6 billion uranium processing plant in Guangdong was scrapped because one thousand people mobilised to denounce environmental and public health risks. Sun Qin, chairman of the CNNC, soberly stated that it was the lack of public acceptance, rather than finance or technology that was holding the industry back. The sector will therefore have to increase the transparency of its activities and establish effective mechanisms for public participation, which will surely require the completion of a comprehensive and much awaited Atomic Energy Law.

Dependence issues: securing and processing uranium, managing nuclear fuel

For some time to come, uranium is bound to remain the fuel that powers nuclear reactors. It is a highly concentrated source of energy that makes up only a minor portion of total generating costs, especially in comparison with coal-fired and liquefied natural gas plants. This offers the advantage of insulating the price of electricity generated from nuclear reactors to fuel price escalation. Yet uranium also gives rise to energy-dependency issues. China possesses at most 1 per cent of the world's recoverable uranium resources, which is, moreover, of bad quality. Given its ambitious civilian nuclear program, China's dependency on uranium imports is bound to increase in the following decades: by 2020, about two-thirds of Chinese uranium is expected to be imported. Beijing will therefore have to develop and consolidate a strong foreign procurement strategy.

In comparison with China's main competitors in Central Asia (Russia, India) or Africa (France, Iran, the USA), China is a relatively new player in the scramble for uranium (Massot and Chen 2013). Nevertheless, the thinly institutionalised, geographically fragmented, post-Fukushima uranium market should allow Beijing to make use of its political and economic assets to secure the flows of the uranium it needs. The Chinese government has taken steps to acquire overseas holdings of uranium in Kazakhstan and Uzbekistan but also in Niger, Namibia and Zimbabwe. While Kazakhstan and Uzbekistan are China's key strategic partners in the Shanghai Cooperation Organisation, Beijing enjoys relatively good relations with uranium-rich African countries. Niger notably looked to China to cut French monopoly over its uranium resources. With regard to Zimbabwe, China was also able to secure its presence by offering a package deal: the rights to mine uranium and diamond in exchange for a nuclear power plant and military hardware.

The principal problems of this strategy essentially lie in a lack of experience in the management of uranium extraction companies and a lack of knowledge of domestic conditions in host countries. In Niger, Tuareg protests against foreign-owned uranium mines are persistent, while miners regularly go on strike to denounce their working conditions. Well aware of the fact that political instabilities will continue to impede smooth flows of uranium from sub-Saharan Africa, China also secured direct uranium imports from stable, uranium-rich democracies such as Australia and Canada.

Other aspects of the fuel cycle also exhibit dependency on foreign firms, one being to enrich and process uranium. Facilities in Western China are well placed to handle stocks imported from Kazakhstan, but given that Beijing seeks to diversify sources, a new facility for the fabrication of fuel rods on the coast would be vital (the scrapped project in Guangdong was expected to fulfil this role). Without the capacity to enrich and process the ore it buys in Asia and Africa, Beijing will still be dependent on foreign firms in order to supply enriched fuel (Stanway 2013).

Conclusions

China hopes to quadruple its GDP by 2020 relative to 2000 and to maintain a strong growth even beyond 2020. To achieve this target, the total primary energy consumption must continue to increase; this will inevitably result in a large gap between China's energy demand and its supply capacity, raising significant energy security concerns. At a certain point, fossil fuels will face resource limitations and current fossil energy technologies will continue to accelerate environmental degradation. In addition to strengthening the country's efforts in energy conservation, the development and use of clean energy (renewables and nuclear) thus seem to be an inevitable choice for diversifying energy supply and ensuring long-term energy security in China. Clean energy technologies are now playing a considerable role, and as these technologies are the only ones able to satisfy energy demand while allowing long-term sustainable development, this is likely to increase in the future. Moreover, the rapid development of manufacturers and developers has shown China's will to increase the role of clean energy at both domestic and international levels.

In China, REN brings much more than just a simple addition of power to the current energy mix. Given the fact that it does not depend on the provision of external resources, REN plays an important role in reducing energy dependence. Seen as necessary by the government, their development is also welcome in the market. Even if some investments were just considered for their profits, they indubitably resulted in the fastest growth of any energy sector. On a global scale, unforeseen opportunities have risen for Chinese players to gain a substantial place on the global market, gaining benefits from careful design technology transfer policies in industries that are not yet fully developed. Finally, while pollution keeps rising in a nation that craves energy, REN also provides an interesting way to focus on the public's concerns while avoiding social unrest. However, the performance of this field – which drives an impressive part of Chinese political decisions – remains tightly bound to not only legal issues, but also to the government's very ability to ensure long-term and stable development.

Despite the fact that it still represents a marginal proportion of the total electricity production capacity, nuclear power has undoubtedly become an important asset in China's quest for diversifying the energy mix and maximising energy independence. To that end, the Chinese government will devote greater attention to the development of the sector, which should increase its nuclear capacity three-fold to 58 gigawatts by 2020, then possibly 200 gigawatts by 2030 and 400 gigawatts by 2050 (World Nuclear Association 2014). Moreover, should China succeed in localising and mastering the construction of nuclear plants, the export of nuclear power could be an important leverage on the international scene. The biggest challenges to China's nuclear development are its fallible security culture, the lack of support from public opinion and the growing dependence on uranium. Nevertheless, China, which has manifestly become the biggest market for nuclear plants, is also at the forefront in terms of nuclear R&D, pursuing reactors powered by thorium, pebble-bed fuel facilities and other Generation IV technologies. Nuclear-powered China is here to stay.

Notes

1 The authors gratefully acknowledge the support of the National Science Foundation of China (Grant no. 71350110537).
2 Here we consider the nature of the resources, the industry's present state and whether or not to include the type of energy production in this chapter.
3 In 2005, the total amount of renewable energy utilised in China was equivalent to about 166 million tons of coal, accounting for 7.5 per cent of total national primary energy consumption (National Development and Reform Commission 2007).

4 Both in the Plenum and in 12th Five-Year Plan.
5 The specific relationship between policy-makers and hydropower companies takes root in early national development plans that are out of the scope of this chapter.
6 The nexus concept recognises the interdependency of water, food and energy security. Because of their role as flood controllers, water reservoir regulators and energy power plants, hydropower dams are both the best illustration of this concept and one of the main management tools.
7 Unofficial translation into English available at: www.ecegp.com/english/knowledge/showknoledge.asp?ID=841.
8 By 'baseline renewable energy sources' we mean sources that are able to produce the baseline for a country energy mix, requiring high predictability and stability. By 'peak production energy sources' we mean energy sources normally used to match national production with consumption during peak consumption hours.
9 Resource curse designates the possible inverse relation between increased dependence on natural resources and economic growth rates. Access to a specific resource can have negative effects on the overall economic development and can lead to counterproductive policies. This can either happen through over-specialisation of the industrial base, the destruction of necessary side activities or inefficient policy design.
10 For mapping completed, under construction and planned projects as of 2014, see the work of Wilson Center: www.wilsoncenter.org/wilsonweekly/cef/china_hydropower/index.html.
11 www.internationalrivers.org/resources/china-overseas-dams-list-3611.
12 It is worth considering that part of those projects, like the 700 megawatts of Zungeru power plant in Nigeria, existed prior to the Chinese international venture but was abandoned because of the lack of resources of the previous international industrial consortium.
13 North Korea, Mongolia, Tajikistan, Afghanistan, Bhutan, Nepal, Russia, Kazakhstan, Pakistan, India, Myanmar, Laos, Vietnam, Kyrgyzstan. If one considers the water basins, the number rises to 18, including Thailand, Bangladesh, Cambodia and Uzbekistan.
14 It is important to note that the issue in this case was far more complex, all projects having now been stopped because of political and foreign investment policy in the country. For more details see: www.internationalrivers.org/campaigns/irrawaddy-myitsone-dam-0.
15 India plans to build a series of water reservoirs in its state of Arunachal Pradesh in order to compensate possible diversion from upstream waters from China. In this case, the sources of conflict are the 36 Chinese projects upstream of the Brahmaputra. The most contentious project is a massive 48,000 megawatts dam slated for the 'great bend' in China before the river swings round into India (over twice the size of the Three Gorges Dam).
16 'Run-on-the-river' projects are the second main category of hydropower generation technology. As opposed to dam projects, the run-on-the-river types neither imply the creation of a reservoir nor allow for the regulation of the river flow. They represent less expensive small-scale decentralised productions source with low environmental impact and no land management potential.
17 Concerning wind power, this chapter only addresses onshore production. While there are important investment schemes and foreign partnership development processes, marine energy and offshore wind are not yet an existing industry in China. Since the future of these branches is still highly dependent on R&D development, this would bring us into prospective studies beyond the scope of this contribution.
18 By suboptimal allocation of public funding.

19 At the time, it represented 1.5 per cent of the nation's power production as reported by the IRENA and GWEC on their report '30 Years of Policies for Wind Energy Lessons from 12 Wind Energy Markets', 2013.

20 It is estimated that over the past few years, an average of 36 RENs (WT) per day have been erected in China.

21 And more broadly 'local content', thereby including all consulting and employment services.

22 Important and well-known consultancy groups publishing data and market reports for the wind energy professionals.

23 This aspect is very well reflected by the decision of Suzlon (the world's fifth REN producer) to sell 75 per cent of its Chinese assets in September 2012, explicating the difficulties of foreign companies.

24 As a very interesting case study, the wind industry is a successful demonstration of technology transfer. What appeared to be a win–win situation for foreign manu-facturers turned out to be a fast track for local companies. High requirements of local content ensured both infrastructure and rapid capacity building, while important market entry barriers prevented international leading companies from dominating the market. This strategy is crucial in understanding how Chinese protectionism contributes to building up independence in power-related issues by reducing the effectiveness of international competitors.

25 This transaction also echoes the difficulties in the global wind market, pushing Vestas to first cut 1,900 and then 3,000 jobs in Europe in 2009 and 2011 respectively.

26 The State Electricity Regulatory Commission indeed identified an increasing number of incidents where quality problems during the construction of turbines (absence of low-voltage ride) lead to damage in the grid equipment.

27 RENs are divided into operating classes corresponding to the maximum annual average wind speed that they are designed to operate in. Class 1: 10 m/s, Class 2: 8.5 m/s, Class 3: 7.5 m/s (we won't enter into subclasses and turbulence considerations).

28 With a total length >50 metres.

29 The 'Three Northern Areas' refers to the northeast, north and northwest areas in China. The region includes 14 provinces/cities: Heilongjiang, Jilin, Liaoning, Beijing, Tianjin, Hebei, Henan, Shanxi, Inner Mongolia, Shaanxi, Gansu, Qinghai, Ningxia and Xinjiang.

30 www.ec2.org.cn/.

31 R&D costs can be deduced at the rate of 150 per cent from income tax. Costs related to training own employees for R&D activities will be deduced from income tax as long as the sum is less than 2.5 per cent of the total base taxable. Accelerated discounting rate for R&D equipment.

32 Free income tax for the first three years and 50 per cent abatement of income tax rates for years 4–6 for PV production firms (standard firm income tax rate in China: 25 per cent).

33 For the moment, no specific development plan or roadmap on CSP has been defined in the 12th Five-Year Plan and there are no specific incentive policies for CSP (standardised power price or feed in-tariff) (ADB 2012).

34 Authors' personal communication.

35 Respectively, 62 per cent and 29 per cent of China's biomass power generation in 2009.

36 NIMBY has not yet affected wind development in China, both due to resource spatial – availability of land – and resource repartition – wind resources are in areas with low population density.

37 Possible emissions of N_2O and SO_2 at four and five times conventional power plants, respectively.

38 Fixed at 20 per cent of coal.
39 A Generation II reactor refers to the class of reactors that began construction in the mid-1960s and make up the bulk of the world's operating commercial reactors. They contrast with earlier prototypes (Generation I).
40 It displays passive safety mechanisms that get triggered by natural forces (such as gravity) in the event of malfunction, making the plant less vulnerable to meltdown, even if electricity shuts down.
41 In this regard, France and the USA have a different approach: while the French government invited caution to be exerted, the USA is providing operational support, training and the handing over of thousands of documents related to the construction of their reactor, conscious that it will take time for China to master its technology and adapt to its safety requirements, hence ensuring its long-term presence in the market.
42 Even for the CPR-1000 reactors, by now considered to be wholly assimilated by the Chinese sector, supplier contracts are retained by at least 85 French SMEs.

Bibliography

Adam, L. (2006) *Wind, Water, Work: Ancient and Medieval Miling Technology.* Leiden: Brill, p. 105.

ADB (2012) 'People's Republic of China: Concentrating Solar Thermal Power Development.' *Technical Assistance Consultant's Report Project.* Available at: www. adb.org/projects/documents/concentrating-solar-thermal-power-development (accessed 20 January 2016).

Ansolabehere, S. and Konisky, D. M. (2009) 'Public Attitudes Toward Construction of New Power Plants.' *Public Opinion Quarterly* 73(3): 566–577.

China Economic Review (2013) 'China Must Play by the Nuclear Rulebook to Win Overseas', *China Economic Review*, August 8. Available at: www.chinaeconomicre view.com/nuclear-industry-CGNPG-Turkey-Areva-Westinghouse-bids (accessed 18 September 2014).

China State Council (2005) *Renewable Energy Law of the People's Republic of China.*

China State Council (2009) *Renewable Energy Law of the People's Republic of China (2009 Amendment).* Available at: www.lehmanlaw.com/resource-centre/laws-and-re gulations/general/renewable-energy-law-of-the-peoples-republic-of-china-amended-2 009.html.

Cho, R. (2011) 'The Push to Dam China's Rivers'. Available at: http://blogs.ei. columbia.edu/2011/05/19/the-push-to-dam-china%E2%80%99s-rivers/ (accessed 26 January 2015).

Collier, P. (2007) 'Economic Causes of Civil Conflict and Their Implications for Policy.' In F. Crocker, O. Hampson and P. Aall (eds), *Leashing the Dogs of War: Conflict Management in a Divided World.* Washington, DC: United States Institute of Peace, pp. 197–218.

Delman, J. and Wang, Y. (2014) 'Fuels for the Future? The Emerging Architecture in China. Liquid Biofuels Market.' In L. Augustin-Jean and B. Alpermann (eds), *The Political Economy of Agro-Foods Markets in China: The Social Construction of the Markets in the Era of Globalization.* Basingstoke: Palgrave Macmillan, pp. 279–306.

Epp, B. (2014) 'China: 2013 Market Statistics and Trends'. *Global Solar Thermal Energy Council* (6 January). Available at: http://solarthermalworld.org/content/china -2013-market-statistics-and-trends (accessed 18 September 2014).

Grimsditch, M. (2012) *China's Investments in Hydropower in the Mekong Region: The Kamchay Hydropower Dam.* Kampot, Cambodia (January).

Hong, J., Li, X. and Zhaojie, C. (2010) 'Life Cycle Assessment of Four Municipal Solid Waste Management Scenarios in China.' *Waste Management* 30(11): 2362–2369.

Kibler, K. M. and Tullos, D. D. (2013) 'Cumulative Biophysical Impact of Small and Large Hydropower Development in Nu River, China.' *Water Resources Research* 49(6): 3104–3118.

Laurent, E. (2013) 'Solar Panels: A Crisis "Made in China".' *Asia Centre special issue.*

Li, J., Cai, F., Qiao, L., Xie, H., Gao, H., Yang, X., Tang, W., Wang, W. and Liu, X. (2012) *2012 China Wind Energy Outlook.* Beijing: Global Wind Energy council.

Liao, C., Jochem, E., Zhang, Y. and Farid, N. R. (2010) 'Wind Power Development and Policies in China.' *Renewable Energy* 35(9): 1879–1886.

Luo, J., Zhan, J., Lin, Y. and Zhao, C. (2014) 'An Equilibrium Analysis of the Land Use Structure in the Yunnan Province, China.' *Frontiers of Earth Science* (DOI: 10.1007/s11707-013-0425-z).

Massot, P. and Chen, Z. (2013) 'China and the Global Uranium Market: Prospects for Peaceful Co-Existence'. *The Scientific World Journal*: 1–11.

Ministry of Environmental Protection (2010a) 'Hanzai Zhishi "liangwu" xiaojian fandan [Drought Results in Extra-Polluting Cuts]', 16 April. *Government of China* (in Chinese). Available at: www.zhb.gov.cn/zhxx/gzdt/201004/t20100416_188275.htm (accessed 18 September 2014)

Ministry of Environmental Protection (2010b) 'Yanzhong ganhan daozhi eryanghualiu paifang liang fandan [Severe Drought Leads to Rebound in Sulphur Dioxide Emissions]', 22 April. *Government of China* (in Chinese). Available at: www.zhb.gov. cn/zhxx/hjyw/201004/t20100422_188549.htm (accessed 18 September 2014).

Nakayama, M. and Maekawa, M. (2013) 'Economic Benefits and Security Implications of Trading Hydropower Through Transboundary Power Grids in Asia.' *International Journal of Water Resources Development* 29(4): 501–513.

National Development and Reform Commission (2007) 'Hédiàn zhōng cháng qī fāzhǎn guīhuà, 2005–2020 [Medium and Long-Term Nuclear Power Development Plan, 2005–2020]' (in Chinese). Available at: www.etiea.cn/data/attachment/123(4).pdf (accessed 18 September 2014).

National Development and Reform Commission (2009) 'Fēnglì fādiàn shàngwǎng diànjià zhèngcè de tōngzhī [Notice on the Improvement of the Pricing Policy of On-Grid Wind Power]' ([No. 1906) (in Chinese). Available at: www.sdpc.gov.cn/ fzgggz/jggl/jggs/200907/t20090727_292846.html (accessed 18 September 2014).

National Development and Reform Commission (2012) *12th Five Year Plan for the Solar Photovoltaic Industry.* Available at: www.americansolarmanufacturing.org/ news-releases/chinas-five-year-plan-for-solar-translation.pdf (accessed 20 January 2016).

Nickum, J. E. (2008) 'The Upstream Superpower: China's International Rivers.' In O. Varis, A. K. Biswas and C. Tortajada (eds), *Management of Transboundary Rivers and Lakes.* Berlin: Springer, pp. 227–244.

Qianzhan Co., Ltd. (2014) *China Solar Thermal Industry Market Prospect and Investment Strategic Planning Report, 2013–2017.* Shenzhen: Forward Business Information Co. Ltd.

Ramana, M. V. and Saikawa, E. (2011) 'Choosing a Standard Reactor: International Competition and Domestic Politics in Chinese Nuclear Policy.' *Energy* 36(12): 6779–6789.

Senate of the United States (2011) 'S.RES.227 RESOLUTION Calling for the Protection of the Mekong River Basin and Increased United States Support for Delaying the Construction of Mainstream Dams Along the Mekong River.' *112th Congress 1st Session.*

Shapiro, J. (2001) *Mao's War Against Nature: Politics and the Environment in Revolutionary China.* Cambridge: Cambridge University Press.

Song, J. (2012) 'China's Solar Thermal Market, Room for Growth', *ChinaDaily,* 20 June. Available at: www.chinadaily.com.cn/bizchina/2012-06/20/content_15515494. htm (accessed 18 September 2014).

Stanway, D. (2013) 'China Struggles to Secure Uranium Supplies After Plant Halted', *Reuters,* 16 July. Available at: www.reuters.com/article/2013/07/16/us-china-nuclea r-uranium-idUSBRE96F17D20130716 (accessed 18 September 2014).

State Council (2012) *China's Energy Policy 2012.* Available at: http://news.xinhuanet. com/english/china/2012-10/24/c_131927649.htm (accessed 18 September 2014).

Tan, X., Zhao, Y., Polycarp, C. and Bai, J. (2013) 'China's Overseas Investments in the Wind and Solar Industries: Trends and Drivers.' *World Resource Institute.* Working Paper (April).

Tu, K. (2012) 'China's Nuclear Crossroads. Carnegie Endowment for International Peace.' *Carnegie* (11 March). Available at: http://carnegieendowment.org/2012/03/ 11/china-s-nuclear-crossroads/a258 (accessed 18 September 2014).

Wang, X. (2013) 'An Analysis of EU-China PV Trade Flows and Domestic Supportive Policies of PV Industry.' *IDDRI.* Available at: www.iddri.org/Publications/Collec tions/Syntheses/PB1613_TS%20MC%20TR_EU%20package_web.pdf (accessed 20 January 2016).

Wang, Y. (2013) 'Drought and Earthquakes Pose "Enormous Risk" to China's Nuclear Plans.' *China Dialogue* (27 February). Available at: www.chinadialogue.net/article/ show/single/en/5746-Drought-and-earthquakes-pose-enormous-risk-to-China-s-nucle ar-plans (accessed 26 January 2015).

Wang, Z., Wang, F., Shi, J. and Li, J. (2007) 'Review and Assessment of China's Renewable Energy Law Implementation.' Technical Report. *Energy Research Institute NDRC,* China.

World Nuclear Association (2014) *Nuclear Power in China.* Available at: www.world nuclear.org/info/Country-Profiles/Countries-A-F/China–Nuclear-Power/ (accessed 18 September 2014).

Wouters, P. and Chen, H. (2013) 'China's "Soft-Path" to Transboundary Water Co-operation Examined in the Light of Two UN Global Water Conventions – Exploring the "Chinese Way".' *Journal of Water Law* (May): 229–247.

WWF and CEFA (2013) *2013 Nián chángjiāng shàngyóu liánhé kě kǎo kě kǎo bàogào* [*Upper Yangtze Joint Evaluation Report*].

Yeh, E. T. and Lewis, J. I. (2004) 'State Power and the Logic of Reform in China's Electricity Sector.' *Pacific Affairs* 77(3): 437–465.

Yu, D. (2012). 'China Waste: the burning issue.' *China Dialogue* (26 January). Available at: www.chinadialogue.net/article/show/single/en/4739-Chinese-waste-the-burning-issue (accessed 26 January 2015).

Zhang, L., Ye, T., Xin, Y., Han, F. and Fan, G. (2010) 'Problems and Measures of Power Grid Accommodating Large-Scale Wind Power.' *Chinese Society for Electrical Engineering.* Proceedings of the CSEE 30(25): 1–9.

Zhang, P., Yang, Y., Tian, Y., Yang, X., Zhang, Y., Zheng, Y. and Wang, L. (2009) 'Bioenergy Industries Development in China: Dilemma and Solution.' *Renewable and Sustainable Energy Reviews* 13(9): 2571–2579.

Zhao, X., Wang, J., Liu, X., Feng, T. and Liu, P. (2012) 'Focus on Situation and Policies for Biomass Power Generation in China.' *Renewable and Sustainable Energy Reviews* 16(6): 3722–3729.

Zhou, J. and Chen, H. (2012) 'Municipal Solid Waste Incineration in China: The Current Practices and Future Challenges.' *International Conference on Future Electrical Power and Energy Systems, Lecture Notes on Information Technology* 9: 346–351.

Zhou, Y. (2013) 'China: The Next Few Years Are Crucial for Nuclear Industry Growth.' *Nuclear Engineering International* 58: 16.

Zhou, Y., Rengifo, C., Chen, P. and Hinze, J. (2011) 'Is China Ready for Its Nuclear Expansion?' *Energy Policy* 39(2): 771–781.

5 Conflicts in the South China Sea

Energy resources, ASEAN and the question of regional stability

François Bafoil

The recent May 2014 confrontation between China and Vietnam in the South China Sea was not an isolated occurrence.[1] Quite the contrary, it belongs to a long series of encounters driven by China, which is seeking to claim these strategically important waters while also protesting against the USA's policy in Asia (Cabestan 2010). At the same time, however, China is pursuing another objective that it considers equally important: peaceful development with its neighbours, a so-called 'peaceful rise' that is intended to help stabilise the 'harmonious world' that China appears determined to foster. China has no interest in triggering the formation of a united front that would oppose its role in the territorial disputes of the South China Sea, particularly since the larger neighbouring powers, including Japan and India, but also the USA, would in all likelihood be part of such an opposition. China's position explains the delicate balancing act that is visible in its foreign policy, which focuses simultaneously on territorial sovereignty and economic cooperation with its Asian neighbours, as well as its Pacific partnerships (Cook 2014; Brink 2014).

Against this background, the role of ASEAN merits close examination. Should the Southeast Asian organisation be considered essentially powerless because of the very principles on which it was founded, which prescribe absolute respect for non-interference in the affairs of third-party countries and in any matter involving national sovereignty? (Beeson 2011; Acharya 2009) Or, alternatively, is the organisation uniquely placed to play a major role, precisely because of the opposing interests of a number of players – on the one hand, the ASEAN member states, and on the other China and, ultimately, Japan, India, and the USA? Indeed, if this solution is the most credible, perhaps it satisfies China's expectations by allowing it not to appear hegemonic, while also meeting the principle of national sovereignty that the other member states so ardently embrace. This study is predicated on this perspective and explores three key aspects of this conflict: economic, political, and legal.

The economic dimension involves the major economic interests that these islands and reefs represent. Firstly, local fishermen are compelled to travel increasingly greater distances to satisfy the growing demand for their seafood. Secondly, the principal monetary value of these islands relates to energy resources that are thought to lie under them, a particularly significant factor for countries that are highly dependent on energy imports.[2] Moreover, no reliable estimate of the total value of the area's energy reserves exists, but figures as high as 7.5 trillion barrels have been mentioned, with half of it situated in the north. Natural gas reserves are allegedly even more extensive (Womack 2011). In May 2011, China's Ocean Development Report estimated that, from 2020, the region's seabed would yield annual revenues of 5.3 trillion Renminbi, the equivalent to approximately 800 billion USD (Van Dyke and Valencia 2000), which would greatly alleviate growing dependency on energy imports. Some drilling was conducted in the 1990s, but the wells are currently blocked by China. Vietnam, whose oil consumption is predicted to triple in the next 10 years, making it a net importer of oil by 2025, is among the countries most interested in the region's sub-surface bounty. In addition, a whole third of the Vietnamese population lives along the coast, and coastal activities account for 50 per cent of the nation's GNP (Raine and Le Mière 2013: 115).

As the conflicts revolve around the question of access to the Malacca Strait, the second dimension explored in this study is geopolitical. The strait is crucial to China's goal of becoming a world-class naval power.[3] Controlling the strait would allow China completely unfettered access to the world's waterways, causing the contested zones to fall under China's jurisdiction, while providing direct access to the Sea of Andaman, and thus to India. If China prevails, 90 per cent of the China Sea would become its sovereign territory, and any ship sailing its waters would require Chinese approval, making the South China Sea tantamount to yet another Chinese lake. As Prem Mahadevan (2013: 40) has noted, the fear is that 'Beijing seems to regard its own Exclusive Economic Zone (EEZ) as an extension of territorial waters, while treating the EEZs of Vietnam and the Philippines as high seas, where no country has exclusive economic rights.' Of course, 8.5 per cent of the energy imported by China imports arrives by sea, 75 per cent of it transiting the Malacca Strait, while more than a third of the world's commercial tonnage and 50 per cent of the world's oil and gas also travels through the region's waters. It is worth recalling that China and India, along with the region's smaller players, account for 65 per cent of the world's population and that the livelihoods of tens of millions of the region's inhabitants depend on coastal and maritime activity (Simon 2012).

Economic and political factors are significant determiners of the third major aspect of the conflict, the tendentious legal environment that surrounds the international property rights. The boundaries of the EEZs (Fig. 5.1) clearly reveal the extensive intersections and overlaps between the various property claims and counter-claims. In addition to islands such as the Spratlys and the Paracels, there are property disputes regarding a number of the region's reefs, which are uninhabitable because they are mostly submerged. The Paracel Islands, which are located between the Chinese Hainan Island and the Vietnamese coast, have been the object of a historic tug-of-war between China and Vietnam. The Spratly Islands (called the Nansha Islands in Chinese), which lie between Palawan, Borneo and South Vietnam, are the focus of a dispute between China, Vietnam, the Philippines and Malaysia. The Sultan of Brunei Darassalam limits his property claims to the coral reefs of Rifleman Bank and Louisa Reefs. The Scarborough Shoal, known in Chinese as Huangyan Island, lies 220 kilometres from the Philippine Province of Zambales and is included in the Exclusive Economic Zones of both the Philippines and China. Indonesia is an appropriate addition to this list due to the boundaries of the ninth Chinese line that touch the territorial waters of the Indonesian island of Natuna (Bonnet 2012). Between boarding and inspections of fishing fleets and nationalist outbursts, these disputes have led to a number of military mobilisations and troop movements, and have also increased the threat of blockades in several locations. To control the islands therefore means to control the South China Sea and, consequently, to control this critical area in terms of global development. It is imperative for a balance to be found, one that not only enables fair and balanced distribution of resources, but that also avoids hegemonic claims by a single regional actor.

The aim of this chapter is to explain ASEAN's role, firstly in terms of the arguments concerning property rights over the islands claimed by China and its neighbours, and secondly in terms of the material interests of the different ASEAN member states. Analysing these questions will then allow for an assessment of the organisation's role as forum and its ability to neutralise confrontations that involve the larger powers – China, the USA, Japan and India – in order to maintain peace in the region. Ironically, the widely acknowledged institutional weakness of ASEAN may in fact be a source of strength, by allowing larger countries to avoid suffering a loss of face. As a consequence, it is possible that instead of weakening ASEAN, the territorial disputes over the South China Sea may actually strengthen its position as an informal arbitrator in the region.

The property problem

History and the law

The arguments on all sides of the conflict have deep historical roots that provide the basis for property claims, at least in theory. However, the parties involved are yet to provide decisive documentation. In reality, a long line of different countries and colonial powers throughout history throughout history has occupied the disputed islands and reefs, but these periods of occupation were not always a matter of formal record. China has cited historical research (Change 2011) to support claims that its earliest occupation dates from the second century BC and lasted for several centuries. There is also some evidence that the Spratly and Paracel Islands functioned as Chinese military garrisons for several centuries. According to Yuan Dynasty (1271–1368) texts, the first reference to Scarborough was first referred to by a renowned astrologer named Guo Shonjing. From the Vietnamese perspective, however, a late fifteenth century atlas is considered to guarantee Vietnamese claims to the Spratly Islands. The Philippines, for their part, cite Spanish recognition of the Paracel Islands in the nineteenth century as support for their own territorial claims. The first evidence that shows that China and the Philippines had attempted to hold serious negotiations dates from the 1930s. At the time, however, neither country published documents to the effect out of fear that imperial Japan would appropriate their claims.[4] Each country ultimately bases its arguments on different periods during which it held the region's islands and rock outcroppings, some of which are recognised by regional actors and in some instances, internationally.[5]

For this reason, all of the parties to these disputes base their claims to territorial expansion of their EEZs on new calculations that are immediately contested by neighbouring countries. Behind these arguments, the debate now centres on the celebrated 'U-shaped line' on Chinese maps of the 1930s (Fig. 5.1). The People's Republic adopted these maps, which were created by Chiang Kai-Shek's government in 1949, without modifying anything. The documents primarily served to define the territorial waters of the newly minted People's Republic of China. They were clearly not used to justify claims to the entire space described by the line, as China has been arguing since 2009. In 1949, China was referring to its 'core interest' (Womack 2011: 37). Indeed, rather than a single continuous line, China began to employ nine discontinuous lines (not 11 as it had earlier been using), beginning in 1953, to establish the boundaries of what was 'Chinese' and, by extension, to define territorial claims within the nation's waters. If these lines had been accepted at the time, non-Chinese watercraft would no longer have been able to navigate freely in the region.

The United Convention of the Law on the Sea (UNCLOS) provides an important method for determining ownership of maritime territory, whereby three criteria are important for doing so: authority over an extended period – China is only able to demonstrate continuity since 1949 (references that are several centuries old, but not completely authenticated); in addition, continuous authority should have lasted for an acceptably extended period, which UNCLOS has established as a half-century, though without specifying the precise meaning of the somewhat arbitrary notion of 'acceptable'; and lastly, other countries support a claim, a remote possibility under the present circumstances (Miyoshi 2012: 5–8).

The question remains as to whether existing legislation regarding an island tend to support Chinese claims and how best to consider the case. An island is defined as a 'naturally formed area of land, surrounded by water which is above water at high tide'.[6] If Scarborough is found to meet this definition, China has precedence over all other claimants and can therefore legitimately claim the island, which would allow China to reign over a vast maritime domain. If this is not the case, however, is it simply a rock outcropping? According to UNCLOS, a marine outcropping is an above-surface area that cannot be sustainably inhabited. Philippine fishermen and their Chinese counterparts can thus benefit from Scarborough, and clearly the same is true of the Spratlys and Paracels for Vietnam, Taiwan, the Philippines, and even the Sultanate of Brunei. The decision is critical for the Philippines, which vividly recalls a 1995 incident involving the Mischief Reef (belonging to the Spratly Islands), when the Chinese occupied the reef before later constructing a naval base.

If China's claim is found to be illegitimate, its claim to domination of the South China Sea would hence be thwarted, but on a much more serious note, such a decision would frustrate China's international naval ambitions, which would be greatly assisted by the ability to create a string of naval bases. As a result, China has deployed considerable ingenuity to prove its claims that these barren stone outcroppings are in fact sustainably inhabitable.[7] Numerous fishing fleets are permanently parked around the atolls, while the construction of an airport and a town is soon to be completed. Plans for tourist development are progressing, and reports from the upper management of CP organisations of the Hainan, Guangdong and Guangxi regions strongly support the activities of fishermen in these three contested areas, where boats have been equipped with sophisticated communications gear that keeps authorities informed of problems that they might encounter in these contested waters (Mahadevan 2013: 46).

ASEAN: a penchant for collective inaction

The most striking feature of the current maritime conflicts in Asia, beyond the various showdowns, which are primarily authored by China, is the absence of a collective response, either by the countries involved or ASEAN. This seeming passivity is consistent with ASEAN's founding principles, which demand absolute respect for the national sovereignty of member states. In cases such as the current maritime property disputes, the organisation is required to seek a consensus that satisfies all parties. This forces member states facing a common threat to jointly find a solution. Under the current circumstances, they can also form limited coalitions, although this appears unlikely to be effective given China's size, or to seek support from major outside players, primarily the USA, Japan and India, for which such conflicts are tailor-made as ways of protesting against Chinese ambitions.

Three groups of countries are involved in the disputes as shown in Table 5.1. The first group involves the coastal countries: the Philippines, Vietnam, Brunei and Malaysia. Next are the countries along the Mekong River: Cambodia, Laos, Myanmar and Thailand. Finally, Indonesia and Singapore constitute a third group.

Limited coalitions/opposing interests

The countries of the first group have little in common other than a shared hostility towards Chinese interference. But appeals to the USA for protection by the

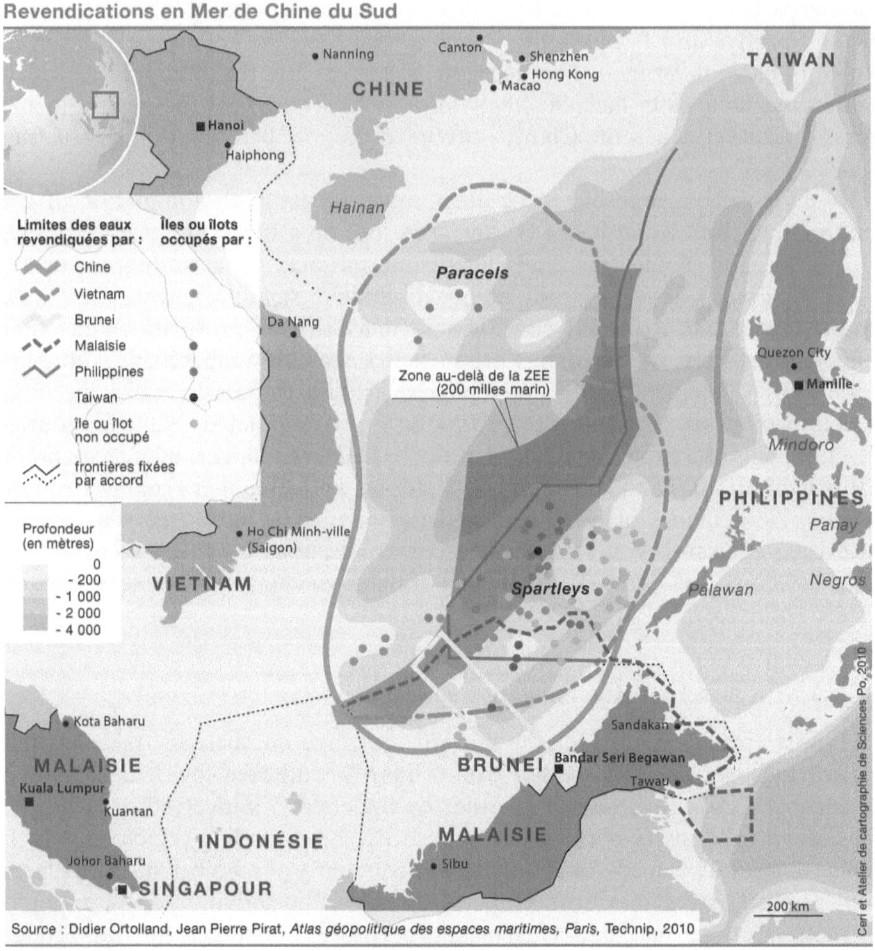

Revendications en Mer de Chine du Sud

Figure 5.1 Maritime zones and claims of countries in the South China Sea
Notes: Thick grey line – China; Dotted and dashed line – Vietnam; Dotted line – Malaysia;
Thin dark grey line – the Philippines.
Source: Prem Mahadevan, 'Maritime Insecurity in East Asia', *Strategic trends 2013*
(Key Developments in Global Affairs) Center for Security Studies, p. 47.

Philippines, and more recently by Vietnam, are not necessarily seen favourably by
Malaysia, and even less enthusiastically by the Sultanate of Brunei, which both
tend to be hostile towards anti-Islamic (or anti-Muslim) policies of the USA. The
Philippines and Malaysia have also developed closer ties to India and the countries
lying to the north, including Japan, South Korea and Taiwan.

On the other hand, although the countries in the second group have often
been in conflict with each other (including Thailand and Cambodia,[8] and to a
lesser extent between Thailand and Myanmar[9]), they uniformly benefit from
the increase in commercial exchange made possible by infrastructure development

Table 5.1 Maritime and non-maritime countries involved in conflict with China

	Internal consistency	*Position concerning China*	*Alternatives and external support*
Maritime countries involved in conflict the Philippines, Vietnam, Brunei, Malaysia	Distended	Conflictual	The USA, Japan, India, South Korea, Taiwan, Russia (Vietnam)
Countries bordering the Mekong River Cambodia, Laos, Myanmar, Thailand	Strong	Cooperative	Weak
Edge countries Indonesia, Singapore	Average	Cooperative	The USA, India, South Korea, Taiwan
ASEAN	Weak	Cooperative	The USA, Japan, India, South Korea

Source: Adapted from David Roy 2005.

as part of the Greater Mekong Sub-region (GMS) (Fau, Kontapane and Taillard 2014). From a financial perspective, each of these countries also needs China, because of its investment in natural gas, rail transportation and hydroelectric dams.[10] Thailand also depends on China because it is its principal market, and it needs the electricity supplied by Laos.[11] Cambodia has become a client-nation due to the fact that China has become its largest foreign investor and principal donor-nation. As a result, China has become one of the major economic forces in Cambodia's remarkable economic recovery in the past decade.[12] Cambodian authorities therefore understandably consider it important to please China, their most powerful ally and supporter. An interesting incident in this context occurred during the ASEAN summit in Phnom Penh in July 2012, when the microphone of the Philippine Minister of Foreign Affairs was 'inadvertently' turned off just as he was on the verge of denouncing Chinese actions in the South China Sea.[13]

Relations between China and Indonesia, the largest ASEAN member state with nearly 300 million inhabitants, and Singapore, the smallest (except for Brunei) with 5.5 million inhabitants, concerning the disputed territories in the South China Sea are overwhelmingly determined by the imperatives of commerce, although tensions have flared a number of times. Agricultural trade between Indonesia and China, as well as commercial exchanges in mining, hydro-electricity and textiles amount to $660 billion. Indonesia provides China with vast amounts of coal and, besides Australia, represents the principal source of China's imports; the two countries combined ensure 60 per cent of Chinese coal imports. In return, China is the source of loans to Indonesia, particularly for financing transportation infrastructure, as well as being the principal supplier of anti-submarine missiles for the Indonesian military. Relations between Singapore and China are generally excellent. In some ways, Singapore is the model of a single-party authoritarian state that has successfully maintained a wholly free-market economy while retaining state-owned companies and exerting extremely strict social control (Ortmann and Thompson 2014). Singapore has often served as China's advocate inside ASEAN and, while a fair number of Singaporean businessmen have hugely greatly benefitted from investments in China since it opened to foreign capital in the 1980s, Lee Kuan Yew's government has also sent numerous experts and specialists to China, particularly to the economic zones (Paix 2010).

Military equipment

The absence of a uniform attitude towards China among ASEAN member states has created asymmetries that are to China's advantage. The organisation is essentially powerless when it comes to influencing this situation, as evidenced by the region's arms race. The ASEAN countries with the most powerful armies – Vietnam, the Philippines and Malaysia – are featherweights compared to the Chinese. It is therefore understandable that these countries should turn to large outside powers for support, chiefly to the USA, but also to India and Japan and, to a lesser extent, Russia. As China is attempting to assert itself as an international superpower, the South China Sea offers excellent terrain for a proxy power struggle with the USA.

As shown in Table 5.2, this has contributed to substantially increased military movements in the region, led by China and which added 1,000 officers to the 10,000 already present in the zone, along with 36 warships to complement a fleet that already contained over 300 and possessed over 10 aircraft (Miyoshi 2012: 12). From 9,000 individuals in 2010, the offshore surveillance force will increase to 15,000 soldiers by 2020. In 2015, the number of planes should exceed 16, with 350 patrol vessels (*China Daily*, June 17, quoted in Sutton, Huang 2011: 71). Over the course of the 2000s, Indonesia and Malaysia responded by respectively increasing arms imports by 84 per cent and 722 per cent (Sheldon 2012: 998). Vietnam spent US$2 billion to purchase a submarine and acquired a fleet of aircraft for an additional $1 billion (Aminul 2013). Most importantly, it has become increasingly diverse in opening itself to other countries, turning towards the USA, with which it concluded an agreement and held joint naval exercises in 2008, and also towards Russia. Indeed, a 2012 treaty with Russia has authorised the Russian navy to utilise the Vietnamese naval base at Cam Ranh Bay in exchange for participation in base maintenance costs. The Philippines has followed this pattern of giant increases in military expenditures, with a military budget that increased in 2013 by a mere 7.6 per cent for modernising the country's navy.

A community of interests

Based on this discussion, is it reasonable to conclude that China is preparing for armed confrontation and intends to use force as a foreign relations tool? Should the slightest sign of discord or unhappiness on the part of China cause

Table 5.2 Military equipment and human resources of the USA, China, Japan and Southeast Asian countries

	Navy personnel	Submarines	Destroyers	Frigates	Coastal patrol craft	Cruisers	Aircraft carriers
USA	333,248	71	61	20	28	22	11
PR China	255,000	71	13	65	211 +	–	1
Japan	45,518	18	29	15	6	2	2
Taiwan	45,000	4	–	22	87	4	–
Vietnam	40,000	2	–	2	62	–	–
Philippines	24,000	–	–	1	63	–	–
Malaysia	14,000	2	–	10	38	–	–

Source: Adapted from IISS, The Military Balance, quoted in Prem Mahadevan, 'Maritime Insecurity in East Asia', *Strategic Trends 2013* (Key Developments in Global Affairs), Center for Security Studies, p. 56.

ASEAN to retreat or capitulate in the name of peace because it has no choice? Since 2002, there have been a number of efforts to create forums for regulating conflicts. The first involved the negotiation between 2005 and 2008 of a three-party agreement between China, the Philippines and Vietnam to verify seismic activity, although no broader, long-term agreement was reached (Chalermpalanupap, 2014). A maritime agency against terrorism and piracy in the Malacca Strait was also jointly established by Indonesia, Malaysia, Singapore and Thailand in an operation called 'Eyes of the Sky'. Some believe that this initiative could evolve into a broader shared forum. Finally, in 2011 an independent body for resolving conflicts in the China Sea was proposed by the Philippines with Vietnam's agreement, but ASEAN failed to ratify the agreement after China rejected it.

In reality, there is a significant cluster of shared interests between China and ASEAN due to a shared historical experience of foreign occupations – by Western powers as well as by Japan – in addition to a shared conviction that in the long run cooperation is more profitable than confrontation (Saul 2013). This explains the fact that China signed the 2002 ASEAN accord, the *Joint Declaration and Code of Conduct (DOC)* and the 2011 agreement with Vietnam, the *Agreement on Basic Principles Guiding Settlement of Sea Issues.* The 2002 declaration followed the 1997 procedure on the conduct of different parties of the China Sea, which China signed calling for the peaceful resolution of conflicts in the South China Sea. This position reiterated in Phnom Penh in 2012, when ASEAN again expressed a collective preference for peaceful settlement.[14] Every observer of the situation considers this agreement to be highly important due to its bilateral nature, imposing ASEAN as a player while also inviting China to exercise moderation and accepting ASEAN's procedures for resolving conflicts.

Clearly, the differences that oppose the Philippines and China are profound, exemplified by the decision made by the former to bring China before an arbitration tribunal regarding Article 287 and Annex 7 of the 1982 UNCLOS convention, which the Philippines demands that China include in its legislation in order to be in conformity with the convention. China refused, however, and responded that the Philippines regularly violates the declaration of Conduct of Parties in the South China Sea. The International Tribunal on the Law of the Sea (ITLOS) ruled that the Philippines' complaint was admissible, however, and named five judges (Sri Lanka, the Netherlands, France, Germany and Ghana) to adjudicate the case. Similar confrontations have not prevented cooperation from continuing within the DOC framework. A Joint Working Group, scheduled to meet four times in 2014, was created to encourage better cooperation. The objective is to develop a strict Code of Conduct, although China declined to participate in favour of a more gradual approach that would nevertheless remain within the framework of the DOC (Chalermpalanupap 2014).

Chinese authorities also decided to participate in numerous meetings that were part of the ASEAN China Summit in 2013. Prime Minister Li even proposed a 'Treaty of Good–Neighbourliness and Friendly Cooperation' between China and the ASEAN countries, albeit without specifying the contents of such a treaty. In July 2013, the Chinese Minister of Foreign Affairs proposed a treaty called the 'Indo-Pacific Treaty' to multilateralise the principles of the TAC (*Treaty of Amity and Cooperation* signed by the ASEAN founders in 1976) in order to combat the destructive effects of the various nationalisms in the region. As Chalermpalanupap (2014) has observed, the basis of the treaty was effective in integrating neighbourliness and friendly cooperation with Russia, while also addressing the possibility of a military alliance, exchanges of military technology, and respect for sovereignty.

Bandwagoning, the balance of power and hedging risks

My analysis of the South China Sea disputes points to several specific observations. The first is simply that ASEAN is able to continue to exist despite the range of internal obstacles relating to the organisation's founding countries. The second observation is that ASEAN's role in the South China Sea region involves a combination of external factors as well as its preference not to be tied to a single partner. The third observation relates to the ongoing and unavoidable presence, and therefore leadership, if not hegemony, of the USA.[15]

Covering risks

The first observation relates to the relative ability of ASEAN to impose itself on the regional chessboard. The situation in the South China Sea conflicts involves the opposition between two kinds of countries: on the one hand, those for whom the balance of power in the region inevitably confers hegemonic power on China; these include the Southeast Asian countries that are limited to 'bandwagoning' (Ross 2005). On the other hand, countries that subscribe to the constructivist school; for these countries, ASEAN provides forums that cause *de facto* norm sharing among the region's players, thus contributing to a balance of power (Acharya 2003). Furthermore, the tug-of-war between bandwagoning, on the one hand, where smaller governments adopt conciliatory positions towards larger players and adhere to their views (thereby abandoning other partnerships) and, on the other hand, the balance of power that allows certain smaller countries to form coalitions that can limit the power of dominant countries to define regional policies. Scholars have labelled this strategy 'hedging' (Roy 2005; Tsung-Yen Chen and Hao Yang 2013).[16]

In turn, these two concepts, bandwagoning and hedging, suggest two analytical perspectives. First, a country that limits its actions to following a dominant country is ultimately at risk in the event of a change of alliance or position on the part of its mentor country. The second point postulates that seeking an alternative to the larger dominant neighbour and risking losing that exclusive relationship can also be risky. The solution appears to be to play the multilateral dimension by developing multiple partnerships that include the dominant neighbour in hopes of reconciling the positions and preferences of each partner. The idea would be to adopt a long-term vision while trying to avoid becoming locked into a single option as well as remaining opportunistic. Evelyn Goh (2007/2008) refers to ASEAN's capacity for 'enmeshing' partners within the organisation's overall strategy as a justification for regional integration. ASEAN's ability to engage every member state in this process relies on a combination of three movements: openness, proposing a shared system of rules, and binding parties to shared objectives (Goh 2007/08). For this reason, although relationships that are based on the patronage of Chinese and American 'bosses' continue to work well for certain Southeast Asian 'client' states, these countries can also serve as intermediaries between the two great powers. Douglas Webber (2010) uses the expression 'honest broker' (with reference to different circumstances) a term that Richard Stubbs (2014) has also adopted.

An honest broker

This ability to function as middleman is not limited to the two largest players in the region, the USA and China; it is also shared by Japan and India. For these two countries, from the north and the west, the China Sea represents a first-order security issue, and ASEAN fully intends to derive an advantage from the recent conflicts between member-states and China. This is especially true of Japan, which is in open competition with China.[17] In 2011, the Japanese Prime Minister Takeaki Matsumoto declared: 'Japan has a great interest in the territorial disputes in the South China Sea because they could have an impact on peace and security in the Asia-Pacific region and they are also closely related to safeguarding the security of maritime traffic' (Raine and Le

Mière 2013: 138). One year later, on 6 April 2012, the Minister of Foreign Affairs declared that 'the South China Sea is the property of the world. Nobody has a unilateral control over it and India is capable enough of safeguarding its interests' (Scott 2013: 55). For Japan, which is in direct conflict with China over Chinese claims to the Senkaku/Diaoyu Islands,[18] ASEAN is a long-term commercial and financial partner.[19] In 2004, the two partners created the Regional Cooperation Agreement on Combating Piracy and Armed Robbery (ReCAAP), which entails the establishment of a centre for exchanging maritime traffic data, particularly in the waters of the Malacca Strait.

In September 2011, Japan assured the Philippines that it supported the latter's territorial claims and proposed to create a 'permanent working group' between the two countries in order to coordinate activities and strengthen military cooperation in the China Sea. In October of the same year, Japanese Prime Minister Shinzo Abe visited Malaysia, Singapore and Indonesia, systematically raising the topic of the South China Sea's islands and waters. Without claiming to enter into the dispute over property rights covering the South China Sea's islands and atolls, Japan has two concerns when it comes to navigation in these international waters. Not only are the Japanese eager to avoid the danger of piracy or disputes that could hinder shipping, they also hope to avoid infringing on international legislation in ways that could in turn impinge on the pre-existing conflict with China over the Senkaku/Diaoyu islands. In this context, preserving the integrity of ASEAN is important, as is supporting the Philippines.

For Indian authorities, the South China Sea represents vital interests (Scott 2013), particularly since the country does not want to limit its influence to the Indian Ocean and hopes to participate fully in Asian affairs (Sheldon 2012: 1010). This notion of 'extended neighbourhood' and its corollary the 'India Pacific Arena' were presented during the ASEAN India Summit in 2012. In addition to investments in transportation,[20] this concept is also economically pertinent in terms of the access Vietnam and Singapore have to energy sources and Indian trade, 55 per cent of which crosses Malacca.

The two oil and gas companies, the Indian public oil and gas company (OGNC) and Petro Vietnam, created an oil drilling and export company that Beijing immediately claimed to be illegal. Finally, in military terms, India benefits from the cooperation with Vietnam and Singapore that took concrete form through joint military exercises, as well as from its access to Vietnamese and Philippine ports, where Indian contingents can be seen floating serenely alongside those of Japan and the USA. In 2012, Thailand, Malaysia, Singapore, Indonesia, Brunei, Vietnam and the Philippines conducted joint naval exercises in the Sea of Andaman, which led to a series of bilateral military accords finally being signed with Singapore, Indonesia, the Philippines and Vietnam. For all of these reasons, India does not intend to see the South China Sea region placed under Chinese authority. India's position can be defined as 'soft containment' of China, the Indian response to the Chinese advances on the Indian Ocean because of the liaison between China and Myanmar.

Leadership...but no power

For some scholars (Min-Hyung 2012; Ba 2009; Ba 2006; Kivimäki 2008), ASEAN has the potential of playing the role of honest broker, meaning that it can play the role of a commercial exchange forum for the larger powers, which at the same time retain their margin of action (Rothstein 1984). It is when the larger powers, in other words the stronger countries, are unable to reach an agreement among themselves that a 'weak' institution such as ASEAN has the potential to play an important role. This is why several scholars have referred to the 'power of the weak'. A number of factors have contributed to ASEAN's ability to function as a middleman in the absence of any regulatory power. The most significant factor has been that the larger players, such as China, Japan and the USA, have tended to neutralise the organisation. Indeed, the USA lost influence in the 2000s by focusing only on terrorism and by blocking the creative initiatives of a monetary fund controlled by Japan, which the USA perceived as competing with the IMF. Today, however, Southeast Asia is a key feature of the Obama administration's foreign policy.

Yet China has fuelled its partners' fears because of its ability to siphon off the vast majority of FDI over the past 20 years, and the permanent menace that it exercises over the region's waters. Indeed, as China's power increased, Japan progressively lost its position in Southeast Asia, while larger Asian players have constantly confronted and attempted to neutralise each other against a backdrop of territorial and historical claims and counter-claims.[21] During the post-Cold War period, an organisation such as ASEAN thus offers the advantage of calming competition and easing the complaints between large players that are seeking forms of agreement that do not force them to compromise their positions. It follows that ASEAN is an outgrowth of the neutralisation of the larger players and of ASEAN's own ambitions to play a key role without any particular member state becoming dominant or behaving in a way that offers the slightest of regional hegemony.

This in turn evokes the fragility of regional balances of power once the stakes transcend the strictly regional level and involve the USA. The USA continues to be the largest consumer market for the region's products. Even if the USA is both economically and financially highly depended on its Chinese partners, it generally balances its dependency by pressuring its debtors not to force themselves into default at the risk of losing everything. Today, as in the past, the way in which the USA wields its enormous influence shapes the preferences of the other players (Beeson 2009). Despite the mess that US foreign policy has caused around the world, it remains the most important global player in terms of values.

However, as China has always been perceived as a threat, it cannot lay claim to a similar role of standard-bearer. China is perceived as riddled with corruption and unlikely to encourage individual liberties, while American values are pervasive in part because of their influence over global development and aid institutions such as the World Bank and the IMF. The influence of these institutions in the region has greatly increased in recent years. Their bilateral strategy in Asia, unlike their policies adopted in Europe, allow them to enjoy highly privileged partnerships with Japan and, through the intermediary of client–patron relationships, to incite certain partners to play the role of spokespersons (in a so-called 'hub-and-spoke' pattern) as a way of more effectively broadening their influence. In the end, the American strategy yields alliances that allow it to enhance its relationships with Japan to the North and Australia to the South, thus effectively containing China.[22] Not to be outdone, ASEAN has created the Trans-Pacific Partnership (TPP), placing the USA at the centre of regional operations, and has also decided to launch a Regional Comprehensive Economic Partnership (RCEP) (Hiebert and Hanlon, 2012). It is worth observing that China is not part of this vast partnership, a clear demonstration of ASEAN's ability to engage in outreach outside of the region.

Conclusion

Ever since ASEAN was first established, there has been a lively debate between those who deny that the organisation's highly constraining principles hinder it from playing a positive influence in the region,[23] and others who recognise the organisation's obvious capacity to utilise a step-by-step process build region-wide strategy (Acharya 1997: 322; Goh 2007/08: 113–157; Stubbs 2014). A close look at the conflicts in the South China Sea tends to corroborate this second perspective. In fact, the constructivist approach appears broadly justified as a means of promoting the ideas and norms that ASEAN is then able to encourage its partners to adopt, particularly China. For this reason, ASEAN seems to function as an effective and important actor in regional diplomacy, as well as to have successfully united a number of 'social networks' via its innumerable forums. As a consequence, some observers have even referred to the 'centrality' of ASEAN (Caballero-Anthony 2014), also noting the fact that the territorial disputes in the South China Sea have elevated the importance of the organisation. It is also true that by not taking any action that might risk perturbing member states, ASEAN keeps them satisfied while also maintaining the conflict at arm's length, thereby remaining suspended between the two superpowers, which in turn show no apparent interest in once again transforming Southeast Asia into a battlefield in the name of their country's individual interests. In this way, ASEAN is proving to be able to remain faithful to its two founding principles: first, the national sovereignty of the member states, in which every country can safely pursue its individualistic strategies without interference from supra-national rules; and, second, the principle of informal relations that supports the maintenance of a peaceful consensus and thus satisfies every party. Together, these two principles justify the organisation's quest, not for limited regional integration of the ten member countries, but for a wider strategy that encompasses external forces while remaining consistent with its commercial strategy and continued economic development. In this sense, it can be argued that 'the ASEAN way' is also an 'Asian way'.

Reformulated in terms of energy resources, the South China Sea conflicts reveal the urgent need for a lasting compromise between China and the USA, as well as Japan and India. This in turns suggests that the solution may involve informal compromise between the two super-powers rather than a formal agreement. Instead of producing an official Sino-American agreement, it is possible that the governance of energy resources in the region should rely on the more flexible and less explicit approaches – in a word, more Asian – such as those espoused and practised by ASEAN.

Notes

1 On May 12, 2014, the Chinese Navy violently repelled the Vietnamese military vessels that had arrived to investigate advances made by Chinese oil explorations near the Triton Islands in the central zone of the China Sea. http://southseacon versations.wordpress.com/2014/05/07/china-vietnam-clash-in-the-paracels-history-st illrhyming-in-the-internet-era/. In 2007 an identical scenario took place, and again in 2011 when Chinese patrol boats cut Vietnamese drilling cables. In June 2012, Vietnam and China launched reconnaissance patrols in the zone, with Vietnam later ruling that any ship sailing in the proximity would be required to have prior authorisation from the Vietnamese authorities. These actions recall the Sino-Vietnamese conflict in 1988 which, under a similar pretext with regard to the Spratly Islands, left 68 Vietnamese dead (other observers have provided estimates of as many as 72 deaths) (Womack 2011: 381).

2 In 2009, China became a net importer of coal and its needs continue to expand. From 2012 to 2011, coal imports increased by 30 per cent and reached 5.2 per cent of consumption. In 2012, China was also the second oil-importing nation in the world (BP 2012), consuming 14 per cent of global oil imports. Finally, in terms of natural gas, China has also been a net importer since 2007, with gas imports representing 29 per cent of demand in 2012 (US EIA 2014).

3 Forty per cent of the crude oil in the Middle East, 80 per cent of it transported by sea, including the Malacca Strait: 85 per cent of oil imports and nearly 15 per cent of coal imports rely on this Strait.

4 On that date, the map shown by the Chinese authorities reflects 11 lines that include the Pratas Islands, the Paracel and Spratly Islands, and the Macclesfield Bank (Miyoshi 2012).

5 This was true only until the French officially claimed possession of the Spratly Islands in 1933 following the colonisation of Cochin China, while also acknowledging, as did the Chinese, that it was Chinese fishermen who resided there, particularly due to the island's proximity to Hainan Province (see Wong KC. 2002: 350–351).

6 UNCLOS, Part 8, *Regime of Islands,* Article 121, cited in Bonnet, 2012, art. cited, p. 6.

7 Also worth noting are 'diplomatic' initiatives revealed by Wikileaks showing that China had since 2006 been pressuring large foreign oil companies – Exxon

Mobil, Chevron, Conoco-Phillips, Japan's Idemitsu, and BP – not to drill in waters claimed by Vietnam in exchange for a promise that these companies would locate concessions on Chinese soil (Sheldon 2012: 1002).

8　The conflict surrounding the temple of Preah Vihar remains unresolved to this day, despite the ruling by the November 2013 International Court of Justice that demanded that Thailand withdraw its troops from the temple (and the 'vicinity'), yet without clearly ruling on the 4.6 square kilometres that constitutes the core of the conflict between the two countries (Khatarya Um 2014).

9　What is at issue here is Burmese immigration into the territories, particularly in the western regions of Thailand, which have led the Thai authorities to decide to conduct violent, forced repatriations (a treatment also applied to Cambodian immigrants).

10　Myanmar is the linchpin of the Trans-Asian Railway project, which is intended to link China to East Asia, crossing the Southeast in the direction of Europe due to the installation of roughly 14,000 km of railway (Vignat 2014). Furthermore, of the 268 hydraulic sites and 35 projected sites listed by the Energy Ministry, the vast majority of these construction projects were granted to Chinese consortiums. This includes the Sweli 1 (600 Mgw capacity), 2 and 3 (480 MGW) dams, those planned for the Nam Ka, a tributary of the Salween, and 6 others on the Nam Lwe River, a tributary of the Mekong, as well as the Le Tasang Dam on the Salween River, the largest in Asia (Simpson 2013).

11　Furthermore, the two countries maintain close relations that are cemented by their shared investments in creating four high-speed railways in Thailand and Myanmar. In addition, Thailand represented ASEAN concerning the China–ASEAN partnership between 2013 and 2015 (Chalermpalanupap 2014: 67).

12　Between 2002 and 2012, China granted $2.1 billion in aid and loans to Cambodia for the development of agriculture and for the construction of 2 000 km of roads and bridges. China currently supports 19 projects for a cost of $1.1 billion. Its trade with Cambodia amounted to $2.72 billion in 2011 (Hill and Menon 2013). According to the Chinese journal *Xinhua*, Chinese investments grew by 62 per cent in 2013 compared to previous years, reaching $427 million, for a total of $9.6 billion since 1994. An article in the *Cambodia Daily* from March 19, 2013 described an iron mining exploration project covering 130 000 hectares in Preah Vihear Province, a concession to Chinese investors with costs initially estimated at $9.6 billion, later raised to $11.2 billion. A 404 km-long railway will link the mine to the autonomous Chinese port south of Koh Kong. It will include 11 stations and employ 20 000 Cambodians and 3 000 Chinese experts. The project also involves a projected 'city of steel' with 50 000 inhabitants.

13　For the first time in its history, the ASEAN summit did not issue a final joint declaration. The Indonesian president responded by calling this patent failure 'utterly irresponsible' and 'extremely disappointing'.

14　*Declaration on the Conduct of Parties in the South China Sea*, 8th ASEAN Summit, Phnom Penh, Cambodia, November 4, 2002 (available at: www.aseansec. org/13163.htm) (accessed 15 June 2014)

15　According to Stubbs, 'Hegemony is the general preponderance, either by force or persuasion, of one state over others either globally or regionally, while leadership is thought of as a process in which one state or group of states in the international system facilitates problem–solving by proposing and helping to execute a course of action in accord with the interests and expectations of a number of other states in the system,' (Stubbs 2014: 14).

16　Royn Tsung Chen and Yoa Yeng use the three concepts 'balance of power', 'bandgwagoning' and 'hedging' to delineate the strategies of ASEAN countries based on consideration of, on the one hand, fears inspired by China in each country and, on the other hand, by their economic expectations. According to the authors, governments that are unafraid of and have low economic expectations with regard to China tend to adopt a 'soft balancing' strategy. Countries that are unafraid but have positive expectations are tempted to follow their northern neighbour; the strategy of 'hedging' is adopted by the others who are either afraid but have positive economic expectations or are unafraid but have no expectations.

17 Competition between China and Japan became intensified during the 2000s in terms of FDI, particularly in the Mekong region (Hidetaka 2010). Furthermore, even if Japan was the number one investor between 2009 and 2011 with $9.6 billion invested, more than the USA at $24.2 billion and China at $10.7 billion, still the rate with which China has caught up is astounding and has steadily accelerated since 2011 (Cook 2014). It will be rightly counter-argued that Japan presents Southeast Asia, particularly through the Asian Development Bank, that it leads and that it wields considerable weight among the 'economic corridors' in the Greater Mekong Sub-region, the support for the fast trains between Kuala Lumpur and Singapore and linking Northeastern India with Northwestern Thailand. In spite of these data, it must be acknowledged that there were 7.3 million Chinese tourists in Southeast Asia (compared to 3.7 million Japanese and 2.8 million Americans) in 2011. In terms of commercial exchanges, China leads its competitors by a vast margin, with $318.6 billion, which it would like to see grow to $1 trillion in 2020.

18 As Ian Storey (2013) has argued, the conflict over the South China Sea islands is equal in importance to the conflict further north over the Senkaku/Diaoyu Islands.

19 In particular via the APT ASEAN + Three (China, Japan, South Korea), the so-called CMI (Chiang Mai Initiative) monetary accord and the Greater Mekong Sub-region.

20 Within the framework of the trilateral association (India, Myanmar and Thailand), India granted a $500 million loan to Myanmar to construct a highway to connect the East–West Corridor of the GMS to the border of Myawaddy Mae Sot. The loan was also intended to finance the modernisation of Myanmar's rail network (see Chalermpalanupap 2014).

21 It should be further noted that ASEAN also lost a great deal in the larger conflicts that have shaken the organisation, including the disastrous management of the monetary swap association, the so-called Chiang Mai Initiative (CMI). In 2006, as Weatherbee (2010) contends, ASEAN failed to support Indonesia's appeal for financial loans and, in 2008, South Korea turned to the Americans instead of to the CMI.

22 '[T]here is plainly a transnational dimension to contemporary processes of governance that may favor some nationally–based elites more than others, without necessarily being unambiguously under the direct control of any of them' (Beeson 2009: 99).

23 For Min-Hyung Kim, ASEAN *'fails to become 'a constructive bridge – builder among major powers in the region...and to provide a genuine alternative way between hegemonic and great middle power leadership,'* Min-Hyung (2012: 112). See also Weatherbee (2010). This is also the position adopted by Malcolm Cook (2014) in his remarks concerning the China ASEAN summit held in May 2014.

Bibliography

Acharya, A. (2009) *Constructing a Security Community in Southeast Asia: ASEAN and the Problem of Regional Order.* London: Routledge.

Acharya, A. (2003) 'Will Asia's Past Be Its Future?' *International Security* 28(3): 150–160.

Acharya, A. (1997) 'Ideas, Identities and Institution–building: From the ASEAN Way to the Asia-Pacific Way?' *The Pacific Review* 10(3): 319–346.

Aminul, M.-K. (2013) 'The South China Sea Disputes: Is High Politics Overtaking?' *Pacific Focus* XXVIII(1): 99–119.

Askew, M. (2010) *Legitimacy Crisis in Thailand*. Chiang Mai: Silkworm Books.

Ba, A.-D. (2009) *Renegotiating East and Southeast Asia: Region, Regionalism, and the Association of Southeast Asian Nations*. Stanford, CA: Stanford University Press.

Ba, A.-D. (2006) 'Who's Socializing Whom? Complex Engagement in Sino–ASEAN Relations.' *The Pacific Review* 19(2): 157–179.

Beeson, M. (2009) 'Hegemonic Transition in East Asia? The Dynamic of Chinese and American Power.' *Review of International Studies* 35: 95–112.

Beeson, M. (2011) *Institutions of the Asia-Pacific, ASEAN, APEC and Beyond*. London: Routledge.

Bonnet, F.-X. (2012) 'Geopolitics of Scarborough Shoal.' *Les Notes de l'Irasec*, 14 (November).

Brink, T. (2014) 'L'accession du capitalisme chinois: l'interdépendance n'interdit pas les tensions.' *Critique Internationale* 63 (April–June): 113–130.

Caballero-Anthony, M. (2014) 'Understanding ASEAN's Centrality: Bases and Prospects in an Evolving Regional Architecture.' *The Pacific Review* 27(4): 563–584.

Cabestan, J.-P. (2010) *La politique internationale de la Chine: Entre intégration et volonté de puissance*. Paris: Les Presses de Sciences Politiques.

Chalermpalanupap, T. (2014), 'ASEAN Managing External Political and Security relations.' *Southeast Asian Affairs*: 53–75. Available at: http://muse.jhu.edu/login? auth=0&type=summary&url=/journals/southeast_asian_affairs/v2014/2014.chalerm palanupap.html (accessed 17 June 2014).

Chang, T. K. (2014) 'China's Claim Sovereignty Over Spratly and Paracel Islands: A Historical and Legal Perspective.' *Case Western Reserve Journal of International Law* 35: 1–21.

Cook, M. (2014) 'Southeast Asia and the Major Powers: Engagement Not Entanglement.' *Southeast Asian Affairs*: 37–52. Available at: http://muse.jhu.edu/ login?auth=0&type=summary&url=/journals/southeast_asian_affairs/v2014/2014.co ok.html (accessed 12 June 2014).

Fau, N., Kontapane, S. and Taillard, C. (2014) *Transnational Dynamics in Southeast Asia, the Greater Mekong Subregion and Malacca Straits Economic Corridors*. Singapore: ISEAS Publishing.

Goh, E. (2007/2008) 'Great Powers and Hierarchical Order in Southeast Asia: Analyzing Regional Security Strategies.' *International Security* 32(3): 113–157.

Hall, H. and Menon, J. (2013) 'Cambodia: Rapid Growth With Weak Institutions.' *Asian Economic Policy Review* 8: 46–65.

Hidetaka, T. (2010) 'The Mekong Region Regional Integration and Political Rivalry Among ASEAN, China, and Japan.' *Asian Perspective* 34(3): 71–111.

Hiebert, M. and Hanlon, L. (2012) 'ASEAN and Partners Launch Regional Comprehensive Economic Partnership', 7 December. Available at: http://csis. org/publication/asean-and-partners-launch-regional-comprehensiveeconomicpartner ship (accessed 19 June 2014).

Kivimäki, T. (2008) 'Power, Interest or Culture: Is There a Paradigm That Explains ASEAN's Political Role Best?' *The Pacific Review* 21(4): 431–450.

Mahadevan, P. (2013) *Maritime Insecurity in East Asia. Strategic Trends (Key Developments in Global Affairs)*. Zurich: Center for Security Studies, pp. 38–59.

Miyoshi, M. (2012) 'China's U-Shaped Line.' *Claim in the South China Sea: Any Validity Under International Law? Ocean Development and International Law* 43(1): 1–17.

Min-Hyung, K. (2012) 'Why Does A Small Power Lead? ASEAN Leadership in Asia–Pacific Regionalism.' *Pacific Focus* XXVII(1) (April): 111–134.

Ortmann, S. and Thompson, M. R. (2014) 'China's Obsession With Singapore: Learning Authoritarian Modernity.' *The Pacific Review* 27(3): 433–455.

Paix, C. (2010) 'Singapore as a Central Place between the West, Asia and China.' In K. Hack, J.-L. Margolin and K. Delaye (eds), *Singapore from Temasek to the 21st Century*. Singapore: NUS Press, pp. 210–242.

Raine, S. and Le Mière, C. (2013) 'Southeast Asia, Between Emerging Great Power Rivalry', Chapter 3, *Adelphi Series, 53*, Routledge, pp. 105–150. Available at: http://dx.doi.org/10.1080/19445571.2013.779491(accessed 18 June 2014).

Rothstein, R. (1984) 'Regime-Creation by a Coalition of the Weak: Lessons From the NIEO and the Integrated Program for Commodities.' *International Studies Quarterly* 28(3): 307–328.

Ross, R. (2005) 'The Geography of the Peace: East Asia in the Twenty First Century', *International Security* 23(4): 81–118.

Roy, D. (2005) 'Southeast Asia and China; Balancing or Bandwagoning?' *Contemporary Southeast Asia* 27: 307–322.

Saul, B. (2013) 'China, Natural Resources, Sovereignty and International Law.' *Asian Studies Review* 37(2): 196–214.

Scott, D. (2013) 'India's Role in the South China Sea: Geopolitics and Geoeconomics in Play.' *India Review* 12: 55.

Simon, W. S. (2012) 'Conflict and Diplomacy in the South China Sea. The View From Washington.' *Asian Survey* 52(6): 995–1018.

Simpson, A. (2013) 'Challenging Hydropower Development in Myanmar (Burma): Cross-Border Activism Under a Regime in Transition.' *The Pacific Review* 26(2): 129–152.

Storey, I. (2013) 'Japan's Maritime Security Interests in Southeast Asia and the South China Sea Dispute.' *Political Sciences* 65(2): 135–156.

Stubbs, R. (2014) 'ASEAN's Leadership in East–Asian Region-Building: Strength in Weakness.' *The Pacific Review* 27(4): 523–541.

Sutton, R. and Chin–Hao, H. (2011) 'China Southeast Relations: Managing Rising Tensions in the South China Sea.' *Comparative Connections* (September).

Tsung-Yen, C. I. and Hao Yang, A. (2013) 'A Harmonized Southeast Asia? Explanatory Typologies of ASEAN Countries' Strategies to the Rise of China.' *The Pacific Review* 26(3): 265–288.

Um, K. (2014) 'Cambodia, the Winds of Change.' *Southeast Asian Affairs*: 97–116. Available at: http://muse.jhu.edu/journals/southeast_asian_affairs/toc/saa.2014.html (accessed 15 June 2014).

Van Dyke, J. M. and Valencia, M. J. (2000) 'How Valid are the South China Sea Claims Under the Law of the Sea Convention?' *Southeast Asian Affairs*: 47–63.

Vignat, E. (2014) 'Shan State in Myanmar's Problematic Nation Building and Regional Integration. Conflict and Development.' In N. Fau, S. Kontapane and C. Taillard (eds), *Transnational Dynamics in Southeast Asia, the Greater Mekong*

Subregion and Malacca Straits Economic Corridors. Singapore: ISEAS Publishing, pp. 190–220.

Weatherbee, D.-E. (2010) *International Relations in Southeast Asia, The Struggle for Autonomy*, 2nd edition. Singapore: ISEAS Publishing.

Webber, D. (2010) 'The Regional Integration That Did Not Happen: Cooperation Without Integration in Early Twenty-First Century East Asia.' *The Pacific Review* 23(3): 313–333.

Womack, B. (2011) 'The Spratlys: From Dangerous ground to Ammle of Discord.' *Contemporary Southeast Asia* 33: 370–387

Wong, K. C. (2002) 'Who Owns the Spratly Islands? The Case of China and Vietnam.' *China Report*: 345–358.

6   China and the Middle East

Moving beyond energy trade

Michal Meidan

China's energy and trade links to the Middle East are rapidly increasing. China is already the region's largest oil consumer and, by the end of this decade, the country is set to surpass the USA as the world's largest oil importer. At the same time, the USA's reliance on Middle Eastern energy is waning thanks to the boom in tight oil production. As a result of the shifting dynamics in energy trade flows, new trade and political linkages are binding China and the Middle East more closely together. In recognition of these changing dynamics, the Chinese leadership has begun beefing up its political dialogues with the Arab League and the Gulf Cooperation Council (GCC), extending its diplomatic outreach to more countries in the region. At the same time, the change in oil flows is contributing to a change in the geopolitical balance of power: as the US commitment to a strong presence in the region comes under question, it is shaping its relations not only with the Middle East, but also with China.

China: a determining factor in global oil markets and for the Middle East

China's importance for in the global oil markets has significantly increased significantly over the past two decades. Ever since the country became a net importer of crude oil in 1996, it has rapidly become a key contributor to the growth in global oil demand growth. Following China's accession to the World Trade Organization (WTO) in 2001, the country's surging economic growth made its unquenchable thirst for oil a determining factor for oil producers and markets. In 2013, China's domestic oil demand reached 9.8 million barrels per day (mbd), almost double its 2003 level, contributing every year throughout that decade to roughly 60 per cent of the growth in global energy demand.

As a result, oil producers globally have become increasingly sensitive to changes in Chinese oil demand and to structural shifts in the Chinese oil industry. Oil producers have begun to look East and shift their focus away from their traditional export markets in Europe and the USA, as Asia Pacific has become the world's biggest growth market for oil. This trend has accelerated with the boom in shale oil production in the USA, which has reduced the US need for imported oil. Nowhere has this shift been felt more significantly than in the Middle East, which has contributed to half of China's oil imports over the past decade and is set to remain a key oil supplier to the country (IEA 2013).

While the growth in Chinese oil demand is expected to slow due to the Chinese government's move to restructure its economy and tackle environmental degradation, overall oil demand in China is still set to rise. The IEA forecasts China's incremental oil consumption at 2.4 million bpd until 2018.[1] And BP estimates that by 2027, China will surpass the USA and become the world's largest oil consumer. With stagnating domestic output at 4.2 million bpd in 2012, the country reached a 57 per cent import dependency ratio. The IEA expects this ratio to reach 80 per cent by 2030, (IEA 2013), whereas BP projects a 76 per cent dependence ratio by 2035 (BP 2013).

China's ravenous demand for oil imports has already skyrocketed from under 2 million bpd in 2002 to 4.8 million bpd in 2012, with roughly two-thirds coming from the Middle East and Africa. The Chinese government has made various attempts to diversify its sources of imported oil, but those have been bearing fruit only gradually. For the better part of the last decade, China relied on a small number of countries for roughly two thirds of its supplies in crude oil: Saudi Arabia, Angola, Iran, Russia, Oman and Sudan. Middle Eastern producers have consistently accounted for the lion's share of Chinese oil imports.

As a result of China's insatiable appetite for crude oil and the USA's declining need for imports, Persian Gulf producers are increasingly focusing on the East, specifically on China, trying to secure as much of the market as possible. China's traditional oil suppliers from the region, notably Saudi Arabia, are now seeing their market share threatened by producers who have been plagued with volatility, including Iran and Iraq, and are therefore now investing heavily in boosting production.

As producers in the Gulf and in Middle Eastern countries more generally come to terms with the changing landscape for oil demand and China's rise as their most significant and enduring customer, the latter country is also reviewing its commercial and diplomatic policies toward the region.

China's rise as the primary global oil importer coincides with two additional trends: the first is the surge in tight oil production following the US shale revolution, which is limiting American demand for imported Middle Eastern oil, and the second is the remaking of China's foreign policy under President Xi Jinping. Both Chinese and Middle Eastern elites are contemplating the impact of a reduced US presence in the Middle East should Washington opt to disengage from the region. While this is not an inevitable outcome, the shift in energy flows clearly coincides with a geopolitical realignment.

China and the Middle East: the pre-oil years

The rapid transformation from net oil exporter in the mid-1990s to the world's largest oil importer in the space of two decades has also shaped

Chinese economic policies and foreign strategies, especially as they relate to the Middle East. However, contrary to many depictions of China's relations with the Middle East, Beijing is no newcomer.

China's relations with the region date back to the Maoist years when ties were shaped by the broader strategic and ideological context of the Cold War and China's foreign policy doctrines (Shichor 1979). China provided aid and logistical support to revolutionary groups even as trade and commercial ties were limited. For China, the Middle East was a welcome observation ground for the balance of powers between the USA and the USSR. These foundations then provided a base upon which ties expanded following Deng Xiaoping's reform and opening up in 1978.

In the mid-1980s as Chinese companies were taking their first steps in overseas investments, the little aid that had previously been provided to the Middle East was abandoned and replaced by nascent commercial ties. Beijing allowed individual workers, and then groups and companies, to undertake labour exports, construction projects and consultation services in foreign countries. Even though it started on an experimental basis, overseas labour contracts gradually became an integral part of China's international economic relations. The opportunity to export manpower and tap into the abundant wealth of oil producers was appealing for Chinese firms. Moreover, overseas construction projects were an opportunity to collaborate with foreign companies and absorb new technologies (Zhang 2000). Host countries in the Middle East welcomed the Chinese workforce not only because of their own shortage in skilled manpower but also because the Chinese were well organized, disciplined and industrious, were not Muslim and did not intend to settle down (Shichor 1999). In the 1980s, most exported Chinese labour left for the Middle East, primarily to Persian Gulf countries, many of which did not yet have diplomatic relations with Beijing.

Many of the firms engaged in construction and services were subsidiaries of China's state-owned oil and gas companies. For example, China Petroleum Engineering and Construction Corporation (CPECC), a subsidiary of China National Petroleum Corporation (CNPC, China's largest state-owned oil and gas company), moved into the Kuwaiti and Iraqi markets in 1983 by competing for subcontracts and small turnkey projects. By 1995, the company won an oil storage reconstruction project in Kuwait valued at $400 million, and this success soon extended to Sudan and Egypt. The Chinese Great Wall Drilling Company (GWDC), another CNPC subsidiary then seized growing drilling business opportunities in Sudan, Egypt, Iran, Qatar, Tunisia, Oman, Libya, United Arab Emirates and Yemen (CNPC 2003). The value of CNPC's overseas oil service contracts increased from $60 million in 1992 to $553 million in 1997 (Xu 2002).

In the late 1980s, about 70,000 Chinese workers were employed in the Middle East, mostly in Iraq and Kuwait. Iraq had become China's primary market for labour export, valued at close to $658 million or nearly 70 per cent of total exports. Iraq had also become an important destination for Chinese construction services, valued at over $670 million (18 per cent of the total construction services overseas), followed by Kuwait, which also absorbed a good deal of Chinese labour and contracted projects, valued at $125 and $364 million respectively. Over time, Chinese labour export was directed to other countries and then on to other regions, and the share of the Middle East in China's total foreign economic activities declined from an annual average of some 56 per cent of the total turnover (of which over 48 per cent represented contracted projects and over 86 per cent labour services) in 1976–1987 to an annual average of nearly 13 per cent in 1998–2004 (Shichor 2006; Zhang 2000).

The second major thrust of Chinese presence in the Middle East came in the early 1980s with weapons sales. China penetrated the regional weapons market due to a growing demand from Egypt, Iraq and Iran. The Iran–Iraq war provided an opportunity for China to supply both belligerents with weapons, since arms sales to the Middle East generated earnings that could be used, at least partly, to supplement China's shrinking defence expenditures. In the 1980s, the Middle East became China's largest military customer, absorbing an average 87 per cent of its total arms sales. In turn, arms became the most important component of China's exports to the Middle East, representing 78 per cent in 1984 and 72 per cent in 1987 (Shichor 2000).

However, from the early 1990s onwards China's arms sales to the Middle East began to decline, almost completely drying up by the end of the decade. While China's total arms sales fell by 60 per cent, its sales to the Middle East fell by over 80 per cent. This decline was due to the arms embargos imposed by the UN and the USA on Iran and Iraq; US efforts to curb military proliferation and the pressure they placed on China as a result; the emergence of alternative arms customers for Chinese sellers closer to home, primarily in South and South East Asia, as well as the emergence of new suppliers to the Middle East; and, finally, to the realization in the Middle East that Chinese arms, although inexpensive, were also inefficient and ineffective under battlefield conditions (ibid.).

Sino-Middle Eastern trade fell to more modest proportions in the early 1990s, mainly consisting of Chinese sales of consumer goods to the region. Middle Eastern capitals were turning to Beijing in an attempt to promote exports in order to even out their trade deficits with China, and the latter's diplomats were therefore looking for a way to increase imports from the Middle East.[2] Faced with demands by Middle Eastern states to balance China's growing trade surplus, the Ministry of Foreign Affairs in Beijing advocated buying more oil, as this was the only commodity that China could conceivably purchase from the Middle East. Ironically, this recommendation was not informed by the recognition of China's looming dependence on oil, but by the basic premise of China's Middle Eastern strategy: to promote commercial and trade ties with the region. However, several plans to acquire oil were rejected by the SDPC, the bureaucratic predecessor of the National Development and Reform Commission, arguing that with global oil prices at low levels, there was no justification to commit to costlier crudes from the Middle East.[3]

The timid beginning of China's oil and gas ties with the Middle East were therefore not the result of a diplomatic strategy to secure energy, but rather a gradual evolution of trade ties. This was spearheaded by Chinese companies and supported by diplomatic interests that were looking to maintain favourable ties with Middle Eastern commercial counterparts, as well as being informed by China's changing ties with the USA and the USSR. Moreover, Middle Eastern capitals provided diplomatic support for Beijing following the Tian'anmen Square protests (1989), when many Western countries imposed embargos or restrictions on China. Oil had become part of the equation because subsidiaries of CNPC were already operating in the region. Moreover, in the 1980s, China periodically bought oil either in order to offset shortfalls in domestic production or to serve as payment for civilian goods and arms (Calabrese 1993).

A cautious evolution beyond oil trade

In the late 1990s, as China's economic growth was gaining pace and domestic oil production was disappointing planners and leaders, oil imports from the Middle East began to surge. However, attempts to deepen ties in the energy sector failed to materialize on both sides.

Chinese economic planners, confronted with the reality that the Middle East was set to become an increasingly important source of imports for China, endeavoured to facilitate oil supplies from the region and upgrade China's refining capacity by using investments from oil producing countries. China's downstream, geared for domestic low-sulphur crude oil, required adaptation in order to be able to process the high sulphur crude from the Middle East. Gulf producers were happy to secure market shares in China by investing in Chinese refineries that would then strengthen their foothold in the market. Negotiations began in the mid-1990s, but Saudi and Kuwaiti attempts to penetrate the Chinese market were stymied by the pricing structure for oil products, which remained administratively set, considerably squeezing the margins of refiners.

Moreover, the competition between China's oil giants for domestic market share eventually excluded foreign investors. Both Sinochem and Sinopec had been in talks with Gulf counterparts for refinery deals, though they were also engaged in a struggle for market shares within China and had different motivations for pursuing talks with producers: Sinochem, created as a trading house with no upstream or downstream investments in China, was hoping that by securing its own sources of imports from the Middle East, it would be able to gain approval for new refineries. As for Sinopec, China's largest state-owned refiner, it wanted to upgrade its refineries and managerial know-how while also securing as much funding as possible. Sinopec was therefore intent on preventing Sinochem from increasing its foothold in the domestic downstream. At the same time, it toyed with the idea of foreign partnerships in the downstream business. Indeed, luring foreign companies with market shares into China has become a strategy for Chinese companies to gain market shares overseas (Li 2004). The lengthy negotiations between Sinopec and Saudi Aramco materialized after over a decade of talks, with a joint venture refinery in China and an investment stake for Sinopec in Saudi Arabia's Rub-al-Khali gas field.

Similarly, Gulf producers welcomed Chinese service companies, yet saw little need to involve them in the upstream. Up until 2003, China had not made appreciable inroads in securing upstream investment deals in its main oil suppliers from the Gulf. Between 1997 and 2002, China's overseas global equity investments were initial and exploratory. Chinese firms had acquired stakes in Kazakhstan, Peru and Sudan, although they had limited opportunities in the Middle Eastern market.

China had signed an agreement to purchase an equity stake in Iraq in 1997, which was pending the lifting of sanctions against the country. CNPC also signed an agreement with two Egyptian companies to form a joint-investment company in 1998, and Chinese firms made their overtures in Iran (Xu 2002; Calabrese 2006).

Ties therefore developed at a gradual pace. Within China, a debate began to emerge regarding the implications of the country's growing dependence on foreign oil, especially on Middle Eastern oil. It seemed natural to planners and companies to shore up ties with the region. However, the reality was far more complex: foreign investments in the Chinese refining industry were greeted with monopolistic resistance from the Chinese national oil companies (NOCs) who were engaged in turf wars for their share of the domestic market. China's administratively set pricing mechanisms also deterred investors. The 'going out' strategy, while hailed enthusiastically, allowed Chinese NOCs to sign many deals and MOUs, only to find the commercial and political realities to be challenging. Upstream investment deals signed with Saddam Hussein's regime were on hold due to the sanctions regime in the country. Even attempts to increase Chinese exports of consumer goods as a means of balancing oil imports were frustrated by practical difficulties, cultural boundaries and local resistance. Nonetheless, in the run up to the second Iraq war in 2003, Beijing had established diplomatic ties with all the regional capitals in the Middle East, and after years of fighting within the region, Beijing has still managed to maintain these diplomatic relations.

At the same time, it was becoming increasingly clear to Chinese decision-makers that it would need to hedge its growing reliance on the Middle East. Not only were regional producers plagued with political instability, but maritime transit routes from the Gulf to China were also rapidly becoming a source of vulnerability for China. The USA's strong strategic presence in the region and maritime transit routes was a major source of concern, as Chinese elites and leading thinkers feared Washington could use its power to cut off supplies to China or manipulate prices. In the time leading up to the Iraq war,

China's oil imports seemed extremely insecure. The political instability plaguing the region and the complex nature of interregional interactions highlighted the difficulties of operating in the Middle East. These trends culminated in the Iraq war in 2003 and were a rude awakening to Chinese planners.

The Middle East and the geopolitics of China's oil dependence

The geopolitics of oil is now widely discussed in China in both official circles and mainstream media. In 2002, as US military intervention in Iraq appeared to be more likely, many in Beijing interpreted this as a ploy to extend the US hegemony on global oil resources (Christoffersen 2005). The notion that supply security was fundamentally geopolitical added urgency to the desire of decision-makers to reduce dependence on Middle Eastern supplies (Liu 2004; Li 2003; Lieberthal and Herberg, 2006).

The subsequent US-led invasion of Iraq highlighted the need for Beijing to mitigate the vulnerabilities associated with relying on the Middle East for oil (Wu 2003). Imports from Iraq contributed to less than 1 per cent of China's total imports, but the prospect of instability in the region from which the country derived half of its imports was daunting (Kong 2010). Concretely, many in China feared that a war in Iraq would result in global price hikes that would translate to higher import costs for China, consequently taking a toll on economic growth (People's Daily Online 2003), as well as the risk that maritime oil supply routes could also be endangered. Furthermore, since the US-led invasion was, according to many analysts (Wu 2003; Yang 2004), all about controlling Iraqi oil, they feared that it would inevitably lead to cancelling China's oil contracts in Iraq. Finally, some estimated that Washington's bid for control over energy resources was not limited to Iraq, but extended to Iran and Saudi Arabia as well, suggesting that China's oil security would soon be beholden to the USA (Wu 2003).

A consistent line of argument within the strategic and foreign policy think tanks in China was that the USA was looking to encircle China and prevent 'China's influence from rising in the region' (Blumenthal 2009) and, after 9/11, they argued that Washington would achieve this by impeding China's access to oil through an increased presence in Central and South Asia. Although Beijing initially showed support for America's war on terror, it became increasingly suspicious of the US presence in Central Asia when the colour revolutions began to unfold (ibid.). Beijing then grew even more concerned as it feared that its overland energy route would slip into turmoil due to political reforms that it thought Washington was orchestrating or at least encouraging (Meidan 2014a).

With an overwhelming sense of insecurity stemming from the country's reliance on the Middle East for oil imports, concerns regarding the slow progress on overland pipeline routes through Central Asia and Russia led Chinese analysts to point to the vulnerability of the security of the sea lines of communication. The fact that over 80 per cent of China's oil imports passed through the Straits of Malacca generated a round of discussions on the 'Malacca dilemma' (Storey 2006). Chinese analysts highlighted the concern that Washington was striving to gain geopolitical pre-eminence in the Strait of Malacca in order to check the rise of China and other powers, as well as control the flow of world energy (Shi 2004).

The inextricable challenges presented by both land imports and seaborne routes gave rise to a school of thought that advocated strengthening ties with producer countries. Former Chinese ambassadors to oil-producing states, as well as scholars at foreign policy research institutes and universities were the main proponents of this strategy (Yang 2001; Wu 2003), a proposition that the oil companies also endorsed as a way of securing additional overseas investment opportunities (Wu, Zhu and Liu 2001). More importantly, some in Beijing hoped that close ties to oil producers would ensure continuity of oil supplies to China in the event of an international crisis. China could offer political and economic benefits to oil-producing states in exchange for access to oil. Politically, China could help its oil-producing allies to reduce the 'Western monopoly' on their oil sector and diversify their foreign partnerships. Some scholars argued that good relations with China could help regimes in the Middle East contain the USA to a certain extent, increase their diplomatic room to manoeuvre and realise diversified security guarantees. They also asserted that China could use its rising power, its involvement in international affairs and a permanent seat on the United Nations Security Council to further the interests of oil-producing states (Wu and Xuan 1999; Yu 2006). Economically, Chinese foreign policy analysts argued that deeper economic interdependence between China and oil-producing countries would help ensure China's access to oil, strengthen trade and investment between China and oil-producing states, open up export markets for China, and make it more difficult for oil exporters to deny China oil in the event of an embargo (Yang 2001; Wu 2003; Zha 2005).

Others endorsed the already common practice of investing in equity oil. The 'going out' strategy now received both theoretical underpinning and strategic endorsement: the basic premise for the need for overseas investment was that acquiring oil through foreign investments and creating overseas 'oil bases' could provide consumers with a more secure and less expensive supply of oil than the international market (Wu and Xuan 1999). The promotion of equity oil resonated with leaders' perceptions that reliance on the international market was a source of vulnerability (Downs 2004) and was also on par with the long-held Maoist principle of energy self-sufficiency. Should China find itself in a situation where it has money but is unable to buy oil, a fear expressed by SDPC officials in the past (Jakobson and Knox 2010), the 2NOCs would be able to send their foreign equity production to China. In addition, some equity oil proponents maintain that barrels of oil produced by Chinese companies abroad would be insulated from fluctuations in world oil prices and can provide the country's consumers with cheaper oil than the international market (Meidan 2014a).

The inescapable allure of the Middle East

The 2002–2003 debate ended when China's energy supplies were disrupted by domestic shortages. These focused the leadership's attention on energy governance rather than on the geopolitics of supply (ibid.). However, China's ever-growing appetite for imported oil and increasing connections with the Middle East generated a flurry of commercial ties and government-to-government connections. Chinese companies, diplomats and top leaders proceeded to develop deeper commercial and political ties with the Middle East, while at the same time trying to cap the share of imports from the region.

For China's oil companies, the leadership's concerns about supply security provided a strategic opportunity to promote their own goals: becoming globally competitive firms, gaining access to knowhow and technologies and acquiring new assets in order to offset declining assets at home (Chen 2008; Ma and Andrews-Speed 2006). In the early stages of China's global oil hunt – the 'going out' policy – the government reluctantly approved overseas investments. However, over time, the idea of creating national enterprises that could be competitive internationally gained ground. The fact that China's oil sector had a powerful patron in government – Zhou Yongkang, the Minister of Land and Resources, who would later become the Public Security czar before falling from grace under Xi Jinping – who also facilitated the pursuit of their interests.

Government attitudes shifted from a reluctant approval to the active support of overseas investments. This then generated support for deal-making through high level visits, extending credit lines to the NOCs, or simplifying approval procedures for investments. Yet the overseas ventures of the NOCs have had limited success when measured in profitability (Yu 2011) or in supply security. By the end of 2010, roughly two-thirds of the $70 billion in overseas investments by Chinese oil companies resulted in loss, and only 100,000 bpd, one twelfth of its overseas output, was shipped back to China,[4] suggesting that these ventures did not afford Beijing greater security. Furthermore, they highlighted the complexities of regional politics that Chinese firms were often unaware of. China's energy ties with Iran and Iraq are a case in point.

China and Iran: friends in need?

The Sino-Iranian trade and energy ties were built on a long history of commercial ties and weapons sales. They have attained significant dimensions over the course of the past decade, as international sanctions on Iran prohibited many Western firms from operating in the country, driving out many of China's competitors. China stepped in as a willing partner at a convenient time: Iran needed access to foreign investment, capital and technology in light of the international sanctions regime. At the same time, Chinese oil and oil service firms became more competitive not only in terms of cost but also in their technological capabilities (Tang 2004). Furthermore, Tehran sought to secure an economic partnership with a rising power that holds a permanent seat on the UN Security Council – a potentially useful lever in countering US and Western pressure.

For China, the rationale for pursuing energy deals with Iran stemmed from a number of factors: Iran is endowed with enormous oil and gas reserves, and it is technically feasible to link Iranian supplies with China by building a connecting line to the Sino-Kazakh pipeline. Energy trading can thus occur overland or by sea, providing an alternative to the vulnerable sea lanes (Calabrese 2006). Furthermore, China's energy officials view Iran as an important commercial opportunity since the upstream in many oil-producing countries in the Middle East is effectively shut to foreign investments (Yang and Yang 2005). More broadly, Beijing views Iran as a regional power and one that will likely emerge from its international isolation and assume a significant position in regional affairs.

China's ties with Iran also have a long history involving trade in goods and weapons. As China proved to be a reliable partner, its oil companies gained access to prize assets. CNPC was granted exploration rights to the giant Azadegan oil field in 2004, then to the Masjed Soleiman (MIS) oil field where it is currently engaged in an oil recovery and extraction project. This was followed by the more impressive deal granted to Sinopec: a $100 billion contract with Iran to annually buy 250 million tonnes of liquefied natural gas for over 30 years, as well as a stake in the Yadavaran oil field in Iran's western Kurdistan province, giving China 50 per cent interest in the field's estimated 17 billion barrel reserve. Yadavaran could eventually become China's biggest oil investment in the Middle East. Moreover, upon commissioning the field, Iran committed to providing China with 150,000 b/d of crude oil at market prices for 25 years (Calabrese 2006; Xu 2000). The following year, CNPC won a bid to develop the Khoudasht oil block in western Iran (Garver 2006; Leverett and Bader 2006; Calabrese 2006).

In June 2006, North Drilling Company (NDC) of Iran and China Oilfield Services Ltd. (COSL), a CNOOC subsidiary, concluded an oil exploration agreement for the management, repair and maintenance of the Alborz semi-floating platform being built by the Iranian Offshore Industries Company. Iran and China also signed a preliminary agreement to construct a gas condensates refinery in Bandar Abbas, aimed at raising the production of gasoline in the refinery's total throughput (Calabrese 2006). Meanwhile, China's main shipbuilding enterprises won contracts to supply Iran with oil tankers. In 2000, the Export-Import Bank of China (Eximbank) agreed to lend $370 million to the state-owned conglomerates, Dalian Shipbuilding Industry Corp and China Shipbuilding Trading Company Ltd., which had been contracted to build five oil tankers for Iran. At the time, this was the largest loan that Eximbank had ever made. Dalian delivered a fourth wide-bodied 300,000-ton VLCC oil tanker to Iran in 2005 (Bloomberg 2005).

Admittedly, Iranian and Chinese officials have flaunted the potential and value of the deals at strategically convenient times, but the actual costs sunk by Chinese firms in the Iranian energy sector are probably lower due to the sanctions regime (Downs and Maloney 2011). Numerous obstacles remain on the commercial, logistical and political fronts. Potential Chinese investors generally perceive market conditions in the Middle East as being worse than those in China (Tang 2005). In the energy sector, it is very difficult for a foreign petroleum company to enter into a mutually attractive arrangement with the Iranian government. The Iranian constitution prohibits production sharing agreements and outright concessions, thus the Iranians have relied on buy-back contracts in which foreign companies act as contractors for the National Iranian Oil Company (NIOC). These companies provide technology and capital and receive a pre-set payment in the form of oil supplies, according to a fee previously agreed upon. Once the foreign firm is paid, all the oil reverts to the NIOC, a fact that has become a stumbling block for many investors. Chinese oil companies have expressed their frustration with the Iranian regulatory regime and the difficulty in making returns on investments (Lee 2007; Downs 2011).

Most of the Chinese upstream investment deals since 2004 are therefore on hold, some due to a reluctance on behalf of Chinese companies to face sanctions in the USA, and others because Chinese and Iranian firms lack the technological knowhow to proceed and need Western partners – all of whom are barred from engaging in Iran (Downs 2011).

More broadly, Chinese companies complain of the lack of support when facing the bureaucratic intricacies of the Iranian market, including local content requirement and legislation that favours domestic firms over foreign investors (Yang and Yang 2005). Iranian firms see the flood of cheap Chinese consumer goods into the Iranian market as a competitive threat (Tang 2005). The UN and unilateral US sanctions on Iran also make financing deals in Iran more complicated and expose foreign financial institutions to potential sanctions on the US market. Maintaining commercial ties with Iran also places a strain on Beijing's bilateral ties with its most important commercial partner – the USA (Shen 2006). Finally, the residual militancy of the Iranian regime 27 years after the revolution has also proven to be a complicating factor for China, which has sought to cultivate the image of 'responsible stakeholder' in regional and global affairs (ibid.).

The geopolitical complexities of operating in the Middle East have also been highlighted in China's investment in Iraq. China's oil relationship with Iraq has gained increased significance since the US-led invasion of Iraq, despite Chinese concerns that it would find itself locked out of the Iraqi oil market.

Iraq 2.0

In October 2006, Iraqi Oil Minister Hussain al-Shahristani embarked on an Asia-Pacific tour to discuss with Chinese energy officials and companies the prospects of reviving the Ahdab oil field deal. Iraqi president Talabani also visited Beijing in 2007 to discuss new forms of cooperation (Ogutcu and Ma 2007).

Despite an initial reluctance to engage in large-scale investments in Iraq, China's NOCs established a foothold in the oil industry of post-war Iraq. CNPC was the first foreign company to renegotiate an upstream contract signed with the regime of Saddam Hussein. The company won three contract bids and won the rights to develop the Rumaila and Halfaya fields with international partners such as BP, TOTAL, Turkish Petroleum and Petronas. In 2008, CNPC also successfully resumed a contract for developing the Al-Ahdab oil field, which it had previously negotiated in 1997 under the pre-war Saddam Hussein regime, making it the only company to have managed such a renegotiation (Jiang and Sinton 2011).

Ironically, despite Chinese concerns regarding the Western domination of the Iraqi upstream after the Iraq war, CNPC became one of the most significant benefactors from the war. Subsequently, in January 2010, CNPC chairman Jiang Jiemin unveiled his plans to place the Middle East at the centre of the company's plans for the decade to follow with regard to upstream growth, focusing on the al-Ahdab, Rumaila and Halfaya fields in Iraq and the North Azadegan field in Iran (China Petroleum News, 2010). Now, a little over a decade after the second Iraq war, Chinese companies have secured a significant foothold in the Iraqi upstream, with roughly 20 per cent of Iraqi oil field projects (Jiang and Sinton 2011), and have also signed deals to more than double the Iraqi exports of crude oil to China. These proposals were launched in 2014 as Baghdad lowered its export prices (Tan and Mackey 2013). Higher crude exports from Iraq are likely to remain a permanent feature, not only in light of production potential from Iraq, but also since part of the upstream contractual structure involves repayment with crude.[5]

While CNPC was successful in navigating Iraqi politics, Sinopec did not fare as well. In June 2009, Sinopec acquired Addax, which allowed the former to add producing assets and reserves to its books. Yet the assets in the Kurdish region of Northern Iraq irked Baghdad and prevented Sinopec's access to the Iraqi market (Caijing 2009).

Is Saudi Arabia still king?

In their largest oil provider, Chinese firms have failed to gain access to the upstream. Despite this, Saudi Arabia has gradually become China's most important energy partner in the Middle East, with ties significantly strengthening after 2001. The tensions in Saudi–US relations following the September 11 attacks prompted Riyadh to diversify its global alliances (Hokayem 2011; Wu 2011; Collins *et al.* 2009). After joining the WTO, China's surge in oil demand propelled it to become the largest source of global growth in oil demand and the single most important export market for Saudi Arabia. Trade and energy cooperation between Saudi Arabia and China therefore rapidly increased as two-way trade reached $74 billion in 2012, more than a tenfold increase over 2002. Moreover, the rising instability in Afghanistan and Pakistan and concerns in China over the potential spillover into Xinjiang provided additional scope for Sino–Saudi cooperation (Hokayem 2011; Collins *et al.* 2009). While there has been limited substantive cooperation on security issues, the confluence of interests has given rise to political support for a bilateral relationship at the highest level: King Abdullah made an official visit to China in January 2006   the first made by a Saudi king to China since the establishment of diplomatic relations between the two countries. President Hu Jintao then reciprocated with visits to Saudi Arabia in April 2006 and February 2009, during which a number of infrastructure investment projects were signed.

A toothless giant in the Middle East?

Just as China's oil trade with the Persian Gulf had increased exponentially, attempts to diversify ties beyond energy failed to gain momentum. Beijing and the Gulf countries launched free trade agreement (FTA) talks in 2004, though these failed to make headway. Negotiations on trade in goods made rapid progress, but talks stumbled on trade in services, and access to China's petrochemical industry remained off limits, frustrating the GCC. By 2009, talks had reached a dead end and GCC member countries decided to pursue their own trade talks with China (Zhang 2009).

Similarly, even though many Middle Eastern capitals tried to encourage Beijing to become more involved in the region's diplomatic and strategic realities, China's interest in taking on a larger security role remains limited. Even though oil supply security is a significant concern for China when it comes to dealing with the Middle East, Beijing has significantly hedged against a potential oil shock through stockpiling and diversifying its import sources (Mitchell 2014). For China, the cost of a deeper strategic involvement in the region is still too high: not only could it undermine China's foreign policy mantra of non-intervention (Duchatel *et al.* 2014), but it could also draw ire to Beijing's policies toward its Muslim minorities. Taking a position in the complex regional questions, such as the Israel–Palestine peace talks or the Iranian nuclear issue, risks hampering Beijing's ties with its top trade partners given their highly divergent interests in these matters.

Indeed, Beijing has stepped up its diplomatic outreach in the region in an attempt to keep abreast of political developments. The need to airlift 36,000 workers during the height of the crisis in Libya highlighted the need to prepare for various types of risk (ibid.). Beijing also recognizes that being a global actor requires both a view and a presence in the top global hotspots. For the past decade, Beijing has therefore been sending a special envoy to the Middle East, and Chinese diplomats have been taking part in P5+1 negotiations on the Iranian nuclear issue. However, a beefed up military presence in the region remains low on the list of strategic priorities. The top diplomatic priorities of Chinese leaders were clearly in the 'near abroad', ranging from potential instability in Central Asia, especially with the looming NATO drawdown from Afghanistan, to shifting dynamics in Northeast Asia.

Shifting paradigms?

There are, however, preliminary signs of a change in China's attitude toward the Middle East due to the shifts underway in the global oil markets and to Beijing's changing views regarding its role in the world following Xi Jinping's accession to power.

The US shale oil boom: exit the USA?

Alongside the US shale gas 'revolution', an output of tight oil produced with the same technology as shale gas began boosting US oil production. Between 2008 and 2012, American oil output increased by 56 per cent. Indeed, the IEA predicts that over the next few years the USA will overtake both Saudi Arabia and Russia to become the world's largest oil producer (IEA 2013).

The market and geopolitical impacts of the increased oil production of the USA are becoming progressively more evident: the growth in US production is acting as a buffer against higher prices due to geopolitical shocks. The abundance of tight oil from the USA has, for example, allowed Washington to pursue export sanctions on Iran as it hoped to take it to the negotiating table (Yergin 2014) without negatively impacting global supplies and prices.

Furthermore, with the USA rapidly emerging as one of the world's largest producers of oil and gas, the 'unconventional revolution' is changing its economic competitiveness and, more importantly, reshaping world energy flows. The USA is gradually importing less oil from the Middle East and West Africa, making more crude oil available to other Asian consumers. The oversupply of oil is also placing downward pressure on global oil prices, allowing China to fill its strategic oil reserves at more favourable prices (Raval 2014). Similarly, since China is set to be the top global oil consumer, the fact that more producers will be competing for market shares will play in Beijing's favour.

Aware of this shift, Middle Eastern leaders are seeking to strengthen ties with China. Over the past year, heads of state, especially from the Persian Gulf, have flocked to China: in September 2013, Bahrain's King Hamad made his first official trip to China since the two countries established diplomatic ties 25 years ago, which was followed by an official visit to Beijing by the Saudi Crown Prince in March 2014. In May, Iranian President Rouhani was in China and the Kuwaiti Prime Minister also made his first official trip to China in ten years (Meidan 2014b).

This is creating new competitive dynamics: Saudi Arabia, China's largest supplier and accounting for over 1 million barrels per day (bpd) of imports, is gradually seeing its share erode as Iraq and Iran establish positions in the Chinese market. With increases in Iraqi output largely due to Chinese investments in exploration and production, crude exports to China have more than doubled, from 300,000 bpd in 2012 to 720,000 bpd in April 2014. Furthermore, even though Iranian exports to China have been weak in recent years due to the sanctions regime, there is the possibility that they could rise in the case of an ease in sanctions. Chinese imports from Iran have already begun to increase, reaching 560,000 bpd in the first quarter of 2014, up from roughly 430,000 bpd throughout 2012 and 2013.

Other Gulf producers are also feeling the pressure: In April 2014, the CNPC gained its first upstream stake in the UAE, along with its Emirati counterpart, the ADNOC. It is Abu Dhabi's first concession deal with China and highlights the UAE's shift away from traditional Western international oil companies and towards Asian partners. Moreover, the CNPC is the ADNOC's sole partner in the venture rather than being part of a consortium, demonstrating the growing appeal of China's oil and gas companies, which can now offer expertise and competitive pricing throughout the energy supply chain (EIU 2014).

From its part, China is reviving its earlier ambitions of supplementing its energy trade with more robust commercial links in a bid to even out its trade deficit and generate new investment opportunities for Chinese firms in the Middle East. During a summit with Arab leaders held in Beijing in June 2014, Beijing announced its goal to increase Chinese non-financial investment in Arab states from $10 billion in 2013 to $60 billion in the upcoming decade. Beijing hopes to expand cooperation in nuclear energy, aerospace technology and new energy.

This renewed interest in the Middle East also dovetails with Chinese President Xi Jinping's new diplomatic doctrine: a creation of a twenty-first-century Silk Road and a Maritime Silk Road. The contours of Xi's foreign policy priorities remain nebulous, but the Silk Roads are already starting to take shape. Both of these modern-day versions of the Silk Road are intended to have key stops in the Middle East: the overland route is set to pass through Iraq and Syria, while the Maritime Silk Road will travel via the Gulf of Aden and the Red Sea. China's vision for these economic belts involves increased trade, achieved in part through major Chinese investments in building up infrastructure such as ports, roads and high-speed railway networks.[6]

China and the GCC, and the Middle East more broadly, appear to have a renewed appetite for taking relations beyond the oil and gas trade. In January 2014, China and the Gulf monarchies resumed the GCC–China Strategic Dialogue following a hiatus of more than two years, caused in part by disagreements over the international response to the conflict in Syria. This reflects a recognition whereby Beijing's growing trade and energy linkages in the Middle East will require more active engagement in regional diplomacy.

This is compounded by the growing concern, both in China and in certain Middle Eastern capitals, that a rapid increase in US tight-oil production will lead to a disengagement from the Middle East (Bellelli 2013). However, this is likely to overstate the extent to which direct oil imports shape US policy towards the region. Indeed, the rising US output will continue to reduce US oil imports from the Middle East and Africa, although Middle Eastern oil supplies have not loomed very large in the overall US petroleum field for some time. After all, even before the growth of tight oil, the Persian Gulf was only providing about 10 per cent of the total US supply. It was not direct US oil imports from the Middle East but rather oil's importance to the global economy and world politics that helped define the strategic interests of the USA (McGregor and Crooks 2013). The Middle East will therefore continue to be an arena of great geopolitical importance, and its oil will be essential to the functioning of the global economy. This implies that the region is likely to remain a central strategic interest for the USA, just as its importance for China continues to increase.

Washington's presence in the Middle East also serves Beijing's interest well: the USA managing security and stability can be vilified if (and when) the region is in turmoil, and ongoing unrest pulls Washington inevitably away from Asia. As a result, even as China expands its ties with the region, Beijing's interests are still best served by maintaining economic ties without participating at a political level. The cost of engagement still far outweighs the benefits.

Certainly, China's commercial interests in the Middle East are also extending to a growing number of countries, drawing Beijing deeper into the complexities of regional geopolitics. Energy ties with Saudi Arabia, Iraq and Iran, combined with the competition originating from other oil producers for shares of the Chinese market, will lead to deeper engagement across the region. However, even though Beijing is beefing up its political dialogues with the Arab League and the Gulf Cooperation Council, there are clear limits to China's engagement in 'the graveyard of empires', at least in the near future.

Notes

1 IEA (2013), *World Energy Outlook*. Paris: OECD.
2 Interview between the author and a Chinese diplomat from the Middle East and North Africa Department, New York, September 2011.
3 Interview with a MFA official, New York, September 2011.
4 Ibid. Other sources may cite different figures, yet they also agree that roughly 10 per cent of China's equity output is actually shipped back to China.
5 Interviews with industry analyst, London, February 2013.
6 Xinhua press agency has a dedicated site to the 21st century silk roads (available at: www.xinhuanet.com/world/newsilkway/index.htm.

Bibliography

Bellelli, J. (2013) *The Shale Gas Revolution in the USA: Global Implications, Options for the EU*. Brussels: Policy Department, EU Directorate General for External Policies.

Bloomberg (2005) 'Hyundai, Daewoo to Share $1 Billion Iran Tanker Order', 9 May.

Blumenthal, D. (2009) 'Concerns With Respect to China's Energy Policy.' In G. B. Collins, A. S. Erickson, L. J. Goldstein and W. S. Murray (eds), *China's Energy Strategy: The Impact on Beijing's Maritime Policies.* Annapolis, MD: Naval Institute Press.

BP (2013) *World Energy Outlook.* Available at: www.bp.com/content/dam/bp/pdf/ Energy-economics/Energy-Outlook/Country_insights_China_2035.pdf (accessed 20 January 2016).

Caijing (2009) 'Sinopec's Access Dilemma in Iraq', 26 October. Available at: http://ma gazine.caijing.com.cn/2009-10-25/110293857.html (accessed 20 January 2016).

Calabrese, J. (1993) 'Peaceful or Dangerous Collaborators? China's Relations With the Gulf Countries'. *Pacific Affairs* 65(4): 471–485.

Calabrese, J. (2004) *Dragon by the Tail: China's Energy Quandary.* Washington, DC: Middle East Institute.

Calabrese, J. (2006) 'China and Iran: Mismatched Partners.' *Jamestown Foundation* Occasional Paper, August.

China Petroleum News (2010) 'Zhongshiyou zhaokai Yilake xiangmu qidong gong-zuohui [CNPC Holds a Work Conference on Launching Iraqi Projects]', *Zhongguo shiyou bao* [*China Petroleum News*], 1 February. Available at: http://news.cnpc.com. cn/system/2010/02/01/001275431.shtml.

Chen, S. (2008) 'Motivations Behind China's Foreign Oil Quest: A Perspective From the Chinese Government and the Oil Companies.' *Journal of Chinese Political Science* 13(1): 79–104.

Christoffersen, G. (2005) 'The Dilemmas of China's Energy Governance: Recen-tralization and Regional Cooperation.' *The China and Eurasia Forum Quarterly* 3(3).

CNPC (2003) *Annual Report.* Beijing: Petrochina.

Collins, G. B., Erickson, A. S., Goldstein, L. J. and Murray, W. S. (eds) (2009) *China's Energy Strategy: The Impact on Beijing's Maritime Policies.* Annapolis, MD: Naval Institute Press.

Downs, E. (2004) 'The Chinese Energy Security Debate.' *The China Quarterly* 177: 21–42.

Downs, E. and Maloney, S. (2011) 'Getting China to Sanction Iran: The Chinese-Iranian Oil Connection.' *Foreign Affairs* (March–April).

Downs, E. (2011) 'China-Gulf Energy Relations.' In B. Wakefield and S. L. Levenstein (eds), *China and the Persian Gulf: Implications for the United States.* Washington, DC: Woodrow Wilson Center.

Duchatel, M., Brauner, O. and Hang, Z. (2014) 'Protecting China's Overseas Interests: The Slow Shift Away From Non-Interference.' *SIPRI* Paper No. 41, June.

Economist Intelligence Unit (EIU) (2014) 'Kuwait Looks to China', 10 June.

Garver, J. W. (2006) *China and Iran: Ancient Partners in a Post-Imperial World.* Washington, DC: University of Washington Press.

Gente, R. (2013) 'Le gaz de schiste chamboule la géopolitique.' *Le Monde Diplomatique*, August.

Guoji Luntan [*International Forum*] 3(1) (February 2001):34–39.

Hokayem, E. (2011) 'Looking East: A Gulf Vision or a Reality?' In B. Wakefield and S. L. Levenstein (eds), *China and the Persian Gulf: Implications for the United States.* Washington, DC: Woodrow Wilson Center.

IEA (2013) *World Energy Outlook.* Paris: OECD.

Jakobson, L. and Knox, D. (2010) 'New Foreign Policy Actors in China.' *SIPRI Policy Paper* No. 26, September. Available at: http://books.sipri.org/files/PP/SIPRIPP26.pdf.

Jiang, J. and Sinton, J. (2011) 'Overseas Investments by Chinese National Oil Companies', *IEA Information Paper*, February. Available at: www.iea.org/publica tions/freepublications/publication/overseas_china.pdf (accessed 20 January 2016).

Kong, B. (2010) *China's International Petroleum Policy*. Santa Barbara, CA: Praeger Security International.

Lee, H. and Shalmon, D. A. (2007) 'Searching for Oil: China's Oil Initiatives in the Middle East', *Kennedy School of Governance Faculty Research Working Paper Series*, March.

Leverett, F. and Bader, J. (2006) 'Managing China-U.S. Energy Competition in the Middle East', *The Washington Quarterly* 29(1): 187–201.

Li, W. (2004) 'Zhongdong Nengyuan yu Zhongguo de Heping Jueqi [Middle East Energy and China's Peaceful Rise].' *Dangdai Shiyou Shi* [*Petroleum and Petrochemical Today*] 12(9): 26–30.

Li, X. (2003) 'Zhongguo nengyuan anquan de bage shijiao [Eight Perspectives on China's Energy Security]', *Liaowang* [*Outlook*], 31 March.

Lieberthal, K. and Herberg, M. (2006) 'China's Search for Energy Security: Implications for US Policy.' *NBR Analysis* 17(1).

Liu, C. (2004) 'Zhongguo shiyou fazhan zhanlue yanjiu [Analysis of China's Oil Development Strategy].' *Shiyou daxue xuebao* [*Petroleum University Journal*] (1).

Ma, X. and Andrews-Speed, P. (2006) 'The Overseas Activities of China's National Oil Companies: Rationale and Outlook.' *Minerals and Energy* 21(1).

McGregor, R. and Crooks, E. 'U.S. Energy Boom Will Not Result in Isolationism', *Financial Times*, 18 November.

Mitchell, J. (2014) 'Asia's Oil Supply: Risks and Pragmatic Remedies', *Chatham House Research Paper*, May. Available at: www.chathamhouse.org/sites/files/chathamhouse/field/field_document/20140506Asia'sOilSupplyMitchell.pdf.

Meidan, M. (2014a) 'The Implications of China's Energy-Import Boom.' *Survival* 5(3): 1–22.

Meidan, M. (2014b) 'China in the Middle East: Where Rising Powers Fear to Tread', *China-U.S. Focus*, 13 June. Availablet at: www.wwv.chinausfocus.com/for eign-policy/china-in-the-middle-east-where-rising-powers-fear-to-tread/.

Ogutcu, M. and Ma, X. (2007) 'Growing Links in Energy and Geopolitics: China and the Middle East.' *Insight Turkey* 9(3).

People's Daily Online (2003) 'Tebie cehua: Duiyi zhanzheng zhongguo jingji sunshi you duoda? [Special Report: How Big Will China's Economic Losses Be as a Result of the Iraq War?]', *People's Daily Online*, 12 June. Available at: www.people.com.cn/GB/jinji/31/179/20030225/930064.html (accessed 10 December 2014).

Raval, A. (2014) 'Growth in Chinese Oil Demand Buoys Brent', *Financial Times*, 21 October.

Shen, D. (2006) 'Iran's Nuclear Ambitions Test China's Wisdom.' *The Washington Quarterly* 29(2): 55–66.

Shi, H. (2004) 'Zhongguo Nengyuan Anquan de Qianzai Weixie: GuoduYilai Maliujia Haixia [Potential Threat to China's Energy Security: Over-Relying Upon the Strait of Malacca]', *Zhongguo Qingnian Bao* [*China Youth Daily*], 15 June. Available at: http://news.xinhuanet.com/world/2004-06/15/content_1526222.htm (accessed 12 December 2014).

Shichor, Y. (1979) *The Middle East in China's Foreign Policy: 1949–1977.* London: Cambridge University Press.

Shichor, Y. (1999) 'China's Economic Relations With the Middle East: New Dimensions.' In P. R. Kumaraswamy (ed.), *China and the Middle East: The Quest for Influence.* New Delhi: Sage Publication, pp. 178–199.

Shichor, Y. (2000) 'Mountains Out of Molehills: Arms Transfers in Sino-Middle Eastern Relations.' *Middle East Review of International Affairs (MERIA)* 4(3): 27–39.

Shichor, Y. (2006) 'Competence and Incompetence: The Political Economy of China's Relations With the Middle East.' *Asian Perspective* 30(4): 39–67.

Storey, I. (2006) 'China's Malacca Dilemma.' *China Brief* 6(8).

Tan, F. and Mackey, P. (2013) 'Iraq Undercuts Saudi to Snare Asian Oil Market Share', *Reuters*, 18 March.

Tang, B. (2004) 'Zhongguo yu Yilang Jingmao Hezuo Qianjing tantao [An Inquiry into the Perspectives of Sino-Iranian Economic and Trade Ties].' *Xiya Feizhou* [*West Asia and Africa*] 5: 58–63.

Wu, B. (2011) 'Strategy and Politics in the Gulf as Seen From China.' In B. Wakefield and S. L. Levenstein (eds), *China and the Persian Gulf: Implications for the United States.* Washington, DC: Woodrow Wilson Center.

Wu, H., Zhu, Y. and Liu, J. (2001) 'Zhongguo shihua tuanti gongsi liyong guoji shiyou ziyuan de zhanlue yu duice yanjiu [Study of Sinopec's Strategy and Countermeasures for Using International Oil Resources].' *Dangdai shiyou shihua* [*Petroleum and Petrochemical Today*] 9(1).

Wu, L. (1997) 'Zhongdong shiyou yu Zhongguo weilai shiyou xuqiu pingheng [Middle Eastern Oil and China's Oil Supply and Demand Equilibrium in the Future].' *Shijie jingji yu zhengzhi* [*World Economics and Politics*] 3: 30–33.

Wu, L. (2003) 'Yilake zhanzheng dui woguo shiyou anquan de yingxing [The Impact of the Iraq War on China's Oil Security].' *Guoji Luntan* [*International Forum*] 5(4) (July): 28–33.

Wu, Q. and Xuan, X. (1999) 'Zhongguo yu Zhongdong de Nengyuan Hezuo [Cooperation Between China and the Middle East on Energy Affairs].' *Zhanlue yu Guanli* [*Strategy and Management*] 2: 49–52.

Xinhua press agency has a dedicated site to the 21st Century silk roads. Available at: www.xinhuanet.com/world/newsilkway/index.htm (accessed 20 January 2016).

Xu, X. (2000) 'China and the Middle East: Cross Investment in the Energy Sector.' *Middle East Policy* 7(3).

Xu, X. (2002) 'China's Oil Strategy Toward the Middle East', taken from the *Post September 11 Update Report: Political, Economic, Social, Cultural, and Religious Trends in the Middle East and the Gulf and their Impact on Energy, Security, and Pricing*, Baker Institute, Rice University, TX. Available at: www.ba kerinstitute.org/publications/PEC911Update_ChinasOilStrategyTowardsMiddleEast 2.pdf (accessed 13 December 2014).

Yang, G. (ed.) (2003) 'Zhongdong feizhou fazhanbaogao 2003–2004: Afuhan zhanzheng hou de Zhongdong wenti, [Report on Developments in the Middle East and Africa 2003–2004: Special Report on Post-Afghan War Middle East].' *Shehui kexue wenxian chubanshe* [*Social Science Documentation Publishing House*] (June).

Yang, S. and Yang, X. (2005) 'Xin Shiqi Zhongguo – Yilang Youqi Hezuo Fengxian Fenxi [An Analysis of the Risks Associated to Sino-Iranian Oil and Gas

Cooperation in the New Period].' *Shijie jingji yu zhengzhi luntan* [*World Economics and Politics Forum*] 5: 100–105.

Yang, Z. (2001) *Zhongya Shiyou yu 21 Shiji de Shiyou Anquan* [*Central Asian Oil and China's Oil Security in the 21st Century*].

Yang, Z. (2004) 'Zhongguo de nengyuan anquan ji zhanlue xuanze [China's Energy Security and Its Strategic Choices].' *Guoji Luntan* [*International Forum*] (May).

Yergin, D. (2014) 'The Global Impact of U.S. Shale', *Project Syndicate*, 8 January.

Yu, H. (2011) 'Oil Majors See Losses in Overseas Investment', *China Daily*, 19 July.

Yu, W. (2006) 'Zhongguo yu Zhongdong Diqu guojia Jingmao Hezuo Xianzhuang he Qianjing Fenxi [The Trade Relations Between China and the Middle East: An Analysis of the Current Situation and Perspectives].' *Shijie Jingji Yanjiu* [*Analysis of International Economics*] 12: 75–80.

Zha, D. (2005) *Zhongguo shiyou anquan de guoji zhengzhi jingjixue fenxi* [*Analysing China's Oil Security From an International Political Economy Perspective*]. Beijing: Dangdai shijie chubanshe [Contemporary World Press].

Zha, D. (2006) 'China's Energy Security: Domestic and International Issues.' *Survival* 48(1): 179–190.

Zhang, J. (2000) 'Zhongguo dui Zhongdong Laowu Shuchu de Qianjing yu Duice [China's Labour Exports to the Middle East: Prospects and Policies].' *Xiya Feizhou* [*West Asia and Africa*] 6: 56–61.

Zhang, M. (2009) *China's Interests in the Gulf – Beyond Economic Relations?* November, National University of Singapore, Perspectives 004. Available at: www. mei.nus.edu.sg/wp-content/uploads/2011/04/MEI-Perspectives-004.pdf.

7 China's new energy haven

Opportunities and challenges in Central Asia

Vanessa Boas and Olga A. Spaiser

Introduction

This chapter will consider China's attempt to ensure access to Central Asian hydrocarbon reserves as a means of fostering its own energy security. Thus far, the Chinese leadership has pursued this objective by investing in exploration, production and development. As it has sought to gain shares in existing projects, it has entered into competition with international energy companies that arrived in the region before China. In the pursuit of these reserves, Beijing has had to heavily invest in new infrastructure so that pipelines can transport Central Asian resources to Xinjiang and develop the country's western province. While trying to make inroads into the region, China has not been able to avoid the gaze of Central Asia's traditional hegemon: Russia. Even if the Shanghai Cooperation Organisation serves as a forum for the coordination of Russian and Chinese interests in Central Asia as well as for the re-shaping of the international system, the relationship between the two does not remain unproblematic. In Russia's eyes, Beijing is a competitor for hydrocarbon supplies and threatens to further dilute Moscow's influence in its near abroad.[1] Moreover, as China's presence grows in the economic sphere, the Kremlin fears that it will acquire political influence at the expense of Russia. Against this background, the Central Asian states continue to resist any form of overt dominance, deepening ties with a range of actors and performing careful balancing acts.

China's voracious appetite for energy and Central Asia's role

China has outpaced the USA as the biggest energy consumer worldwide. To date, its energy supply has been ensured mostly by sea transport: 80 per cent of its oil import has been channelled through the Strait of Malacca (Hayward 2009), which connects the Indian Ocean with the Pacific Ocean. A blockade of this strategic sea lane would cut China off from its energy supply, a scenario that could occur in case of conflict with Washington or Taiwan. For this reason, China explores alternative transport and supply routes, where Central Asia comes to the fore as part of the solution. Any assessment of Sino-Central Asian relations, however, would be incomplete without taking into account the issue of Xinjiang, China's northwest province that borders three of five Central Asian republics. The stabilisation of this agitated province needs to be seen as a crucial part of Beijing's energy security programme.

China's energy needs

Chinese economic growth – averaging 9.2 per cent in the years between 1989 and 2013 – has been at the root of increased global demand in energy resources (Taborda 2014). While China was once believed to have had abundant supplies, its growth spurt has resulted in consumption rates exceeding those of domestic production. As a consequence, the Chinese leadership has increasingly sought to ensure access to energy reserves beyond its own borders. According to forecasts, this trend will only be further accentuated, with 65 per cent of total Chinese crude oil consumption being met by exports by 2015 (Malhotra 2012).

Given its unquenchable thirst for oil and growing appetite for gas, it is no surprise that China has set its gaze on the neighbouring Central Asian region, which harbours substantial natural resource deposits.[2] Since negotiations with Russia over a gas agreement were strenuous until the final signing of a 30-year deal in May 2014 (Pizzi 2014), Beijing had consistently been eager to build up energy links with the whole Eurasian region, and continues to do so. By virtue of their vast proven energy reserves, Kazakhstan and Turkmenistan are the most palatable of the five Central Asian countries, although Uzbekistan is growing in importance thanks to Chinese investment in the hydrocarbon sector. In spite of their relative dearth of hydrocarbons, relations were also deepened with Tajikistan and Kyrgyzstan, as these countries offer Beijing a range of opportunities in the raw material and trade sector. Central Asia currently provides China with over 10 per cent of its oil and gas imports (Mariani 2013: 10) (Table 7.1).

In addition, Central Asia is a lucrative market for Chinese goods and thus a window of opportunity for enhanced trade relations. Because of these

Table 7.1 Proven oil and gas reserves in Central Asian countries

Country	Proven oil reserves	Proven gas reserves
Kazakhstan	30.00 billion bbls	85 trillion cubic feet
Turkmenistan	0.60 billion bbls	265 trillion cubic feet
Uzbekistan	0.59 billion bbls	65 trillion cubic feet
Kyrgyzstan	0.04 billion bbls	0.200 trillion cubic feet
Tajikistan	0.01 billion bbls	0.200 trillion cubic feet

Source: US Energy Information Agency 2012.

important commercial ties, China's status in the region is deemed to have reached that of an economic superpower, especially following the visit of President Xi to Kazakhstan, Kyrgyzstan, Turkmenistan and Uzbekistan in September 2013 (*The Economist* 2013). This *tour* resulted in a pledge of US $48 billion in investments, many of these targeting the hydrocarbon sector (Mashrab 2013). In this context, it must be noted that Central Asia figures very low in the list of China's trade partners, not reaching more than 1.4 per cent in China's total trade balance.[3] Nevertheless, the region is of strategic and economic importance for the aforementioned power as it allows it to pursue a number of development and stability-related objectives.

Central Asia's potential

Originally considered a forgotten region, Central Asia was closed off to the outside world during Soviet times with limited foreign penetration in the hydrocarbon field. With the fall of the Soviet Union, there was sudden demand for foreign investors and technologies to develop Kazakhstan's deep, high-pressure fields that had previously been deemed too complex by the Kremlin. As production rates skyrocketed in the 1990s and new fields were discovered, Chinese attention was diverted to Kazakhstan as a potential major overland oil provider for its burgeoning economy. A similar development can be noted in the case of Turkmenistan, where increased openness to foreign investors through the change in Turkmen leadership at the end of 2006 coincided with growing Chinese gas requirements (US Energy Information Administration 2012b).

From a geographical and infrastructural point of view, Central Asia is a complex region despite its vast hydrocarbon potential and increased willingness to cooperate with foreigners. In addition to being landlocked, its railroad, pipeline and road networks are still underdeveloped. Moreover, as all major export routes were traditionally directed towards Russia, the Central Asian states were limited in their saleable quantities, while also being subject to Russian pressure. To a certain extent, investments from abroad have allowed the countries in question to reduce Russia's leverage in this sphere. Nevertheless, despite improvements in all countries since independence, transport links within and beyond the region remain weak.[4]

With regard to its international standing, Chinese plans to diversify its trade routes away from the sea to land by relying on railway networks crossing Central Asia are likely to raise the region's profile (Mashrab 2013). While transportation via Central Asia is more costly than the use of sea routes, it is also less time consuming, which could be of particular interest to Western manufacturers based in China. Moreover, the building of such links through Eurasia would also increase the interconnectedness of the region itself, with railroads being built between Kyrgyzstan and Uzbekistan. Given the high dependence of the region on trade in relation to its GDP, this is likely to simultaneously benefit its economic development (Mogilevskii 2012).

The Chinese have supported the building of pipelines and have backed infrastructural projects in the region by granting credits to the governments in power. As Central Asia is reachable by land from China and not vulnerable to sea blockades, it enjoys a key position in China's energy security strategy. As a result, Beijing is currently working on expanding the capacity of the existing Central Asia–China gas pipeline (crossing Turkmenistan, Uzbekistan, Kazakhstan and Xinjiang) and, from 2016 onwards, extending it to the region's poorest and least stable countries – Kyrgyzstan and Tajikistan – through the so-called 'Line D' (Lelyveld 2014). There are similar plans to expand sections of the Kazakhstan–China oil pipeline in order to increase its capacity from 12 to 20 million mt/year by 2015 (Mason 2013).[5]

It is worth noting that Sino–Central Asian energy relations also fulfil another purpose: Beijing seeks to smoothen out economic disparities between Eastern and Western China by pumping hydrocarbons into the impoverished Xinjiang province. Hoping to boost development and thus allay anti-Chinese sentiment, relations with Central Asia also have a stabilising value that should quell Islamic fundamentalist currents among the Uighur minority. Finally, by winning over Central Asian governments as partners, Beijing hopes to extinguish any flames of Turkic solidarity, thereby cutting off external support to the Uighurs.

Central Asia's benefits

As China has taken on the role of the major investor and trade partner of the Central Asian states, it is deemed to have joined the 'New Great Game'. While this originally took place between the British Empire and the Russian Empire in the nineteenth century over influence in Central Asia, China is part of the contemporary scramble for the region's hydrocarbon reserves and pipeline construction process. It thus competes with the European Union, the USA, Russia and, to some extent, India, Iran and Turkey for access to oil and gas fields in these energy-rich states. The benefits that Central Asian states can reap from the presence of a range of actors in the region cannot be denied. The competition between powers has the potential to enable the governments concerned to obtain inflated prices for their strategic goods and to work with the highest bidder. This reality greatly contrasts the depictions of a Great Game[6] in which weak and voiceless states are solely pawns on the chessboard of great powers.

As a partner, China relishes several advantages over other actors due to the nature of Chinese foreign and internal policy. In contrast to the West, China cooperates freely with dictatorships such as Turkmenistan and does not advocate regime change. Nor is it accountable to domestic constituencies who challenge its behaviour on normative grounds, as is the case with public opinion in Western democracies. In line with the Chinese pursuit of economic rights over individual rights, the governments of the region obtain loans and investments without being subject to critique for their human rights records. Another advantage of collaborating with China is that its investments are equally long-term and generally not subject to change, thereby providing a solid foundation for partnerships. In addition, Chinese state-owned banks freely make capital available and the Chinese government has generally been willing to overpay for hydrocarbons in order to ensure future access to reserves. At the same time, they provide much-needed loans for development services such as the building of bridges and refining complexes, along with their investments in the oil and gas sector, giving them a clear advantage over competitors (Malhotra 2012).

The Kremlin, in contrast, is perceived as being willing to use a range of tactics to coerce the region's leadership into agreeing to initiatives that help Russia re-assert itself in its near abroad. Such endeavours generally entail a loss of Central Asian sovereignty and are thus regarded critically. As its former patron, Moscow is deemed to meddle in the domestic politics of CIS members and to try and shape local outcomes behind the scenes. Moreover, the annexation of Crimea and the Ukraine crisis in 2014 greatly contributed to mistrust in Central Asia and gave fuel to anti-Russian rhetoric (Spaiser 2014).

The issues surrounding cooperation with the West are of a different nature. While the USA and the EU set restrictive conditions and timelines for reform, cooperation with the two can provide domestic legitimacy and prestige as well as capital, technology and know-how. Neither China nor Russia can compete on this level at present, even if China's potential to have economic, and consequently political, influence is increasing exponentially thanks to its investments in the region.

In their relations with foreign powers, Central Asian states must calculate the cost and the benefits of engagement. While relations with the West may be risky because governments can be subject to spontaneous and unpredictable attacks for their human rights records, it can speed up the exploration and development process in the hydrocarbon field. In the case of Russia, cultural and linguistic similarities facilitate cooperation but also engender paternalistic behaviour on behalf of their former patron. China, in comparison, is a colossal beast with an alien culture and therefore brings with it the benefit of almost limitless appetite for investment.

Ruining the appetite

Ensuring energy security via close cooperation with Central Asia, however, bears some serious challenges for China. Firstly, Beijing has to find a way to deal with Russia, which still regards the region as its main sphere of influence. Secondly, it has to face political instability – brought by interstate tensions, regular blockades of transport links and a lack of regional cooperation – in Kyrgyzstan and potentially in other Central Asian countries as well. Thirdly, China's economic and demographic weight triggers fears among local populations and results in anti-Chinese sentiments, which could hamper Beijing's long-term energy projects.

Who else is sitting at the table? Dealing with Russia

As the traditional hegemon of the region, Russia seeks to safeguard its privileged position in the realm of politics and security, while equally ensuring access to Central Asian energy reserves, traditionally considered to be situated in its exclusive sphere of influence. However, Moscow has increasingly had to make way for other actors due to it losing its geopolitical weight internationally in the wake of the dissolution of the Soviet Union.

In the field of politics, Moscow prefers leaders who take its regional interests into account and are willing to participate in its integration projects. The Russian presence in Central Asia is considered indicative of its international standing and thus a matter of great importance if the country is to regain influence in global affairs. At the same time, the Kremlin seeks to protect its local minorities and safeguard the Russian language. Much to Moscow's dismay, its mission has been greatly complicated by local nation-building processes that emphasise the rebirth of indigenous culture and greater traditionalism. The Russian language is increasingly being replaced by local languages, especially among nationalist circles. Reinforced resistance also stems from the perception that Moscow is indifferent to local interests and to a partnership between equals.[7]

From a security point of view, Central Asia is deemed to be a buffer against the instability emanating from Afghanistan and Pakistan. Moreover, drug flows and illegal migration tend to pass through Central Asia on their way to Russia's long permeable state borders (Jonson 1998). The flourishing of Islamic groups, which could wreak havoc in Russia, is also at the forefront of concerns and actively addressed through cooperation.

When the five states of Central Asia are compared in terms of their attitudes towards Russia, the differences are glaring. Kazakhstan is generally considered to be the most pro-Russian country and is involved in a number of Russian security and economic initiatives, such as the Collective Security Treaty Organisation (CSTO)[8] and the Eurasian Economic Community (EurAsEC).[9] Turkmenistan and Uzbekistan, however, have stayed on the sidelines, largely stressing their sovereignty and autonomy. As the poorest countries in the region, Kyrgyzstan and Tajikistan are in a weaker position and thus have been more prone to cooperating with their former patron.

Russia and energy access

Trade between Central Asia and Russia has centred on energy products and raw materials, which is a legacy of the USSR (Sinitsina 2012). While Russia does not generally consume the fossil fuels it imports from the region, it relies on these to meet its commitments to Ukraine and Western Europe. Despite the presence of competing actors, Russia still strives to ensure its access to Central Asian energy supplies in order to meet its obligations to third states. Since independence, the Kremlin has consistently tried to prevent Central

Asian markets from leaving its grasp by pushing out foreign investors in an attempt to safeguard its position on the local market and ensure that its position as a global energy exporter not be jeopardised (Jonson 1998: 60). In pursuit of its objectives, it controlled energy flows out of the region via its pipeline network and placed substantial political pressure on the states in question. This resulted in a struggle between the Central Asian elite and their former patron as greater independence and room for manoeuvre was pursued.

The Kazakh case clearly illustrates this. The government has tried to diversify its export routes away from Russia by relying on trans-Caspian tankers and rail, as well as the pipeline to China. In spite of this, the relationship between the two states remains close in the sphere of energy: Kazakh oil still largely transits Russian territory, while Moscow and Astana continue to jointly develop a number of oil fields in the Caspian Sea. A similar trend can also be noted in Turkmenistan. While Russia was historically Turkmenistan's primary export and transit market, gas-related disputes over prices and volumes have pushed Turkmenistan to diversify its list of customers. Regardless of the fact that Russia started paying market prices for Turkmen gas from 2008 onwards, China and Iran have increasingly been pursued as partners in order to raise production levels and ensure Turkmen sovereignty (US Energy Information Administration 2012b).

In Uzbekistan, the scenario is more mixed. Uzbekistan currently sends over half of its gas to Russia following a long-term agreement between Gazprom and Uzbekneftgaz, which gives it a leading position in the Uzbek energy market (Karimov 2012). Furthermore, both Gazprom and Lukoil are involved in the hydrocarbon industry and are seeking to raise production output while searching for more deposits (Karimov 2012). At the same time, Uzbekistan has also sought to attract other companies with greater investment potential and newer technologies (Karimov 2012).

The picture in Kyrgyzstan and Tajikistan remains one of dependence. Russia currently enjoys a monopoly over Kyrgyz oil imports with purchases pegged at 1,150.000 tonnes a year (Rickleton 2013a). Potential tariff manipulations by Russia can be employed to put pressure on the leadership, hence why Kyrgyzstan is currently developing its own oil refinery (Rickleton 2013a). Tajikistan has also sought to reduce its reliance on Russian oil by enhancing its own hydrocarbon potential. Exploration is now being carried out by Gazprom and the Canadian Tethys Petroleum, though other actors have been invited to invest in exploration and drilling (Sodiqov 2012).

Integration in the post-Soviet space

While Russia observes the advances of other actors, it seeks to consolidate its position in the region by promoting a number of new integration initiatives such as the Customs Union and the Eurasian Union.[10] Although these projects are economic, their driving force is largely political as Russia attempts to restrict the expansion of other actors in the economic realm. However, these projects do not represent major threats to China since the proposed Customs Union is not likely to have noticeable effects in terms of trade diversion from China due to the volumes of trade flows being so large. In reality, no country in the Customs Union can compete with cheap Chinese goods that can be found at bazaars all over the region.

Despite these initiatives, the Central Asian states remain divided over Russia's integration projects, particularly in the aftermath of the annexation of Crimea (Balci 2014). This stems from the fact that rising trade flows and greater economic development will inevitably go hand in hand with greater Russian influence in the region. Sovereignty remains a major concern in all the CIS states, even if economic needs coax them into considering integration. In the case of Kazakhstan, integration is associated with pragmatism. As Kazakhstan sees Russia as a strategic partner that it cannot oppose, it seeks to benefit from the protection the arrangement guarantees its market from cheap Chinese goods as well as the consequent strengthening of its negotiating position vis-à-vis China. At the same time, Astana enjoys facilitated access to the Russian market by virtue of its membership, leading to a rise in export rates and greater FDI. However, the Kazakhs stress that integration is to be economic rather than political, as well as on a basis of equal partnership so that it does not threaten Kazakh sovereignty; this way Kazakhstan can be protected from the Chinese advance without falling under Russia's influence.

Negotiations for Tajikistan and Kyrgyzstan are underway, though both countries remain reluctant given the uncertain benefits membership may bring for their already weak and non-competitive economies. Moreover, both fear that their (re-)exports of Chinese products could be hit by a project of the like, as favourable Chinese tariffs would be cut (BBC 2014). For Moscow, their membership marks the reaffirmation of its position of dominance in the region and also serves as a barrier to Afghan drug flows as the Customs Union border could be extended.

Uzbekistan and Turkmenistan, however, are the most critical of Russia-led integration. Both underline their sovereignty and refuse to be in a Russia-dominated organisation, although Uzbekistan has toyed with the idea of joining, though without finalising its decision (Sadykov 2013). This behavioural inconsistency has been characteristic of the leadership in Tashkent, which has already left a number of CIS integration initiatives such as the CSTO and the EurAsEC, whereas Ashgabat has categorically defended its neutrality since 1995.

Sino–Russian competition over trade

Central Asian states can partly afford to remain aloof around Russia due to the increased presence of China in their trade balances. When the data regarded compares years 2000 and 2010, it is evident that Russia has been pushed aside by China.[11] In terms of imports, volumes substantially increased for both actors, although the total share of China in the trade balance has risen by four (in formal trade alone), whereas that of Russia has remained almost constant. In terms of exports, the decline of Russia's share is visible, with China gaining greater importance in the field of exports (Tables 7.2 and 7.3).

Substantial differences also exist between countries: China holds first place in Kyrgyzstan in overall trade turnover, enjoying a 51.6 per cent share and surpassing Russia's 17.6 per cent share (European Commission 2012). In the Kazakh case, China is also ahead of Russia, even if the difference between the shares of both countries is not as large (23.6 per cent against 19.1 per cent) (European Commission 2012). While Russia is ahead of China in Uzbekistan (18.7 per cent against 17.5 per cent), their positions are very close (European Commission 2012). In Turkmenistan, China's position is unquestioned, with a 45.3 per cent share compared to 6.8 per cent for Russia, whereas China also leads with a 36.9 per cent share in Tajikistan, compared to 14.3 per cent for Russia (European Commission 2012). These data illustrate the Chinese dominance in the trade sector of the Central Asian states, a trend that is likely to be accentuated with the growth of the Chinese economy.

Common interests and points of contention

Despite Russia's trade losses, Sino–Russian relations have deepened via Central Asia, a region both actors view with great interest. Shared objectives such as ensuring Central Asia's stability currently form the backbone of their relationship. As regional conflict and terrorism would directly threaten their investments and energy projects, these are issues both actors try to tackle in order to prevent economic losses. Moreover, as both Moscow and Beijing have witnessed terrorist attacks in recent years, potential extremist networks based in Central Asia are to be dismantled through cooperation. Forestalling the spread of extremism and terrorism to their territories also figures highly on the national security agendas of both countries. A final factor uniting the two is the growing resentment both governments feel concerning the mounting presence of the

Table 7.2 Central Asian imports from abroad

	Billion US$ in 2000	*% of total imports in 2000*	*Billion US$ in 2010*	*% of total imports in 2010*
Russia	3.1	27.2	17.2	27.3
European Union	2.2	19.0	11.1	17.5
China	0.28	2.4	6.8	10.7

Source: Mogilevskii (2012).

Table 7.3 Central Asian exports going abroad

	Billions US$ in 2000	% of total exports in 2000	Billions US$ in 2010	% of total exports in 2010
European Union	3.7	23.8	31.9	37.7
Russia	3.6	23.3	13.8	16.4
China	0.7	4.8	12.4	14.6

Source: Mogilevskii (2012).

USA in the region, as well as its dominance internationally. As the war in Afghanistan resulted in deepening political and military ties between the USA and Central Asian states, a joint effort was deemed necessary in order to counter Washington's growing influence. Despite these overlapping interests, the integration projects of both actors remain undoubtedly a bone of contention as both hurt each other's trade interests. While China desires to open the markets of Central Asia and facilitate the flow of investments, Russia wishes to further protect its own economic interests and remain competitive. To date, constraints posed by Russia in Central Asia have not elicited a strong reaction from China as it vows to pursue a peaceful rise. However, as Beijing increasingly overshadows Moscow economically in the region and gains leverage vis-à-vis Russia in its bilateral ties in the sphere of energy, greater outspokenness may be noted on behalf of Beijing, both within Central Asia and beyond.

It is all about energy

Central Asia is not the only bone of contention that has previously marred Sino–Russian cooperation. Despite the fact that Russia has immense hydrocarbon reserves and China an insatiable appetite, Sino–Russian energy relations have frequently stalled over price disputes and general mistrust. While Moscow has feared a loss of sovereignty concerning China, due to the potential growing dependence on China to fill its coffers, Beijing has expressed doubts about Russia's reliability as an energy provider given its penchant for using deliveries as a political weapon (Indeo 2013).[12]

The economic downturn in Europe and the 'Shale Revolution' in the USA have pushed Russia to consider the benefits of working with its larger neighbour in light of the consequent decline in demand for Russian gas (Indeo 2013). Recognising the urgency of locking China into long-term energy deals, Rosneft agreed to both a 25-year supply deal of 360 million tonnes of crude with a value of US$270 billion (Bierman 2013) and subsequently a 10-year deal for the delivery of approximately 200,000 barrels of crude oil a day (Yep 2013), totalling US$85 in the year 2013 (Dyomkin 2013). In addition to an agreement targeting the capacity expansion of the Eastern Siberia–Pacific Ocean (ESPO) pipeline,[13] Novatek also sold China a 20 per cent stake in a

US$20 billion Russian liquefied natural gas (LNG) project, which will ensure the export of three million metric tonnes of LNG (Bierman 2013).

Moreover, in May 2014, Moscow and Beijing finally concluded a long-awaited US$400 gas deal after a decade of negotiations. After years of stagnation, it can be presumed that the sanctions imposed by the USA and Europe on Russia's annexation of Crimea were the catalyst towards a final deal over prices, since the Kremlin was eager to seek an alternative market for its gas and oil. Only three months later, Vladimir Putin launched the construction of what will become one of the largest gas pipelines in the world, stretching nearly 2500 miles from Russia's Far East to China. Thus, the new Sino–Russian rapprochement in the energy sphere is also aimed at creating a powerful energy bloc in the East and thus diminishing Western dominance in general and the dollar dominance on world's energy market in particular (Pizzi 2014).

In light of the above-mentioned commitments, Russia will have to develop further fields in East Siberia, which China has offered to support through joint ventures (Bierman 2013). While these deals allow Moscow to reduce its debt and make further acquisitions, it also substantially reduces Russia's leverage vis-à-vis China (Kogtev 2013). As the Kremlin depends on energy exports to keep its economy afloat, it will increasingly have to placate China, both bilaterally and regionally, in Central Asia once greater supplies start flowing east.

Political instability leaving a sour taste? Dealing with Central Asian vulnerabilities

Increased ties between Central Asia and China have implications when it comes to the stability and economic development of the region as a whole. While regional infrastructural projects allow China to reinforce its presence locally and in its agitated Xinjiang province, it also enables this landlocked area to integrate into the world economy, thereby improving the prospects of Central Asian exports to reach beyond their own confines. However, the numerous interstate, intrastate and cross-national tensions in the region could impose limits on Beijing's ambitions.

As far as interstate relations are concerned, regional cooperation between the Central Asian republics could bring clear benefits. Nevertheless, the leaders of Central Asia generally struggle to agree on regional initiatives and are loath to renounce any sovereignty. This reticence to partake in regional cooperation endeavours will potentially hamper Chinese regional infrastructural projects, aiming to facilitate trade within and beyond the region (Rickleton 2013b).

Severe visa regimes, regular transport connection blockades and border closings have recently increased in the region and further heat up existing disputes over borders, water, energy and arable land (Rotar 2013; Rotar 2014). In addition, the domestic political situations in most Central Asian countries do not provide comforting stability prospects. Kyrgyzstan has been a theatre for two popular uprisings, in 2005 and 2010, with the outcome of two presidents being ousted from power. A violent clash between Uzbeks and Kyrgyz in Southern Kyrgyzstan in June 2010 with 420 casualties and around 400,000 displaced persons (International Crisis Group 2012: 2) revealed strong ethnic tensions, a widening north–south divide and the weakness of the central government in Bishkek. Uzbekistan is regularly involved in legal disputes with foreign investors and arbitrary policies against international companies and has thus earned the reputation of being a highly unreliable and instable business partner. Tashkent ranks 146th out of 189 countries in the 2014 *Doing Business Report* issued by the World Bank (EBRD 2013). These perspectives are additionally worsened by the unresolved, albeit urgent, question of succession of aging Islam Karimov who has been in power since the independence in 1991. Voices from both inside and outside predicting power struggle disputes, including civil war scenarios, are on the rise (e.g. Hale 2013; Goble 2013). With regard to Tajikistan, the only Central Asian republic that experienced a civil war, from 1992 to 1997, yet it is still the most vulnerable country of the region with its 1,400-km-long common and porous border with Afghanistan to offer an attractive terrain for local and external insurgencies (International Crisis Group 2011).

In theory, the aforementioned Line D pipeline has conflict-resolution potential as it passes through Uzbekistan to Kyrgyzstan and Tajikistan and thereby ties these actors to each other. Due to energy and territorial disputes, the latter two countries chiefly have tense relations with Tashkent. This has resulted in Uzbekistan cutting off natural gas supplies to both countries in 2010, while shootings have also taken place between Uzbek and Kyrgyz troops (Rickleton 2013b). At the same time, with the construction of Line D, the region's poorest countries Tajikistan and Kyrgyzstan are given the opportunity of buying gas from sources other than Uzbekistan and Kazakhstan, which have often been accused of abusing their position of dominance over the markets of the former two countries.

Being devoured? Dealing with neo-colonialist fears

Apart from regional hostility, anti-Chinese sentiment can also undermine Beijing's projects for regional integration, and Central Asians fear a loss of sovereignty vis-à-vis China (Beshimov 2013). Indeed, the extreme power differential between the small, young and fragile republics and the demographic and economic giant triggers fears of being absorbed by China. A look at the demographic evolution of China best illustrates the reason for anxiety: the Chinese population grows by more than 15 million persons each year. This annual demographic surplus is almost equivalent to the entire population of Kazakhstan (Peyrouse 2012: 103). In addition, the flooding of Central Asian markets by cheap Chinese products is perceived as a threat to their fragile and nascent domestic industries. Similarly, the presence of Chinese workers in an environment marked by increasing pauperisation and high unemployment rates adds to anti-Chinese sentiments. There have thus been reports of attacks against Chinese companies and their workers, especially in Kyrgyzstan.[14] While such attacks have taken place (Ng 2014), it is yet to be determined whether such resentment has not been instrumentalised by Moscow through the Russian language media in order to stem the Chinese advance (Mashrab 2013). This also serves nationalist parties and movements that actively exploit such fears and try to catch votes with slogans calling for an end of foreign occupation and neo-colonialism from East and West. Even though there are no systematic anti-Chinese sentiments in the region (Rotar 2012), Chinese business actors are aware of the risks for their investments and projects. The head of the Chinese Chamber of Commerce in Kyrgyzstan, Li Deming, did not hesitate to call Kyrgyzstan a 'mine field for investors' and urged Chinese investors to pay more attention to the 'country's political background, nationalism ..., social conflict ... as well as distribution of interests' (Deming 2012).

Meeting the challenges

Challenges include political instability, Russia's dominant role in Central Asia and Central Asian fears of being overrun by China. Beijing is aware of them and the possible impediments that it needs, either to overcome or to extenuate in order to achieve its goal of enhanced energy security and the development of alternative supply routes. To this end, it has developed a wide range of bilateral and multilateral instruments.

Ensuring energy supply and gaining trust: Bilateral agreements, tailored to the specific contexts of each republic

Following its ambition to secure long-term supplies of energy from Central Asia to China, Beijing relies on a vital network of intergovernmental ties and bilateral engagements. At the same time, it takes the challenge of seriously enhancing its soft power in order to build trust in its northwestern neighbourhood. However, trade opportunities, infrastructural projects, investments and loans are the most effective instruments at China's disposal to attract Central Asians.

China's Grip on Kazakh Oil

Kazakhstan has the eleventh largest proven oil reserves in the world, coming second in the CIS only to Russia (The Business Year 2012). In this context, it must be noted that this Post-Soviet state's oil production has substantially increased since the fall of the Soviet Union,[15] owing to the closing down of inefficient factories and growing investments in the oil sector from abroad (Overland *et al.* 2009: 118).[16] Hydrocarbon exports have allowed Astana to boast one of the fastest growing economies in the CIS, while simultaneously rendering it resource-dependent and a potential victim of the Dutch disease.[17]

Kazakhstan wishes to diversify its export routes away from the Russian transit network, which gives it too much power over the price and amount of oil sales (Kalicki 2001). Oil swaps with Iran, participation in the Caspian Pipeline Consortium (CPC)[18] and the Kazakhstan–China crude oil pipeline all served this end (Overland *et al.* 2009: 119). China is seen as a key partner in this process due to its keen interest in ensuring the copious flow of hydro-carbons to its borders and its willingness to pay higher prices than other actors (Overland *et al.* 2009: 117).

Beijing's relations with Kazakhstan date back to the latter's independence in 1991, yet only began to reach strategic importance for both towards the end of the 1990s when Chinese demand for oil and raw materials began to grow (Paramonov *et al.* 2011c). As this happened the Chinese government began offering credits for financing projects in the hydrocarbon and raw material sector while simultaneously seeking to gain access to iron, copper and titanium eposits. The Chinese weight in Kazakhstan's economy was further reinforced by the acquisition of stakes in companies involved in the hydro-carbon sector (Demytrie 2010). Examples of this are the purchase of PetroKazakhstan by CNPC for US\$4.2 billion in 2005 (Mariani 2013: 10) and the acquisition of a 50 per cent stake in MangistauMunaiGas in 2009 (Burashev 2009).

Chinese ownership of Kazakh reserves currently remains low, while its presence is greater in production, which requires less advanced technology (Overland *et al.* 2009: 125). However, as China recently gained a much-desired stake of 8.33 per cent in Kashagan – the largest developing Kazakh oil field[19] – China's presence on the Kazakh energy market is likely to increase (Jarosiewicz 2013). At present, relations centre on a 2,798-km-long pipeline – with a capacity of 10 million tonnes a year – that allows for the transportation of crude oil from Western Kazakhstan to China's Xinjiang province (Hydrocarbons Technology). This endeavour is estimated to cost US \$3 billion and is backed by China's largest state-owned oil company, the National Petroleum Corporation, and Kazakhstan's counterpart, KazMunai-Gaz (Hydrocarbons Technology). Moreover, Kazakhstan has benefited from China's loans-for-oil scheme, which has allowed it to inject large sums of money into the economy in return for long-term oil delivery commitments (Malhotra 2012).

Kazakhstan's forecasted oil output of 100 million tonnes by 2015 and Chinese growing oil demand lay down the foundation of a fruitful partnership in the years to come (Hydrocarbons Technology). Nevertheless, awareness of China's population size and land scarcity keeps distrust alive in Kazakh society, as illustrated by anti-land sale demonstrations (Paxton 2011). However, as Kazakhstan has become increasingly assertive in its relations with external powers and has not hesitated to disrespect property rights in the name of national interest (Overland *et al.* 2009: 133), it is unlikely that one actor will dominate the country and, instead, multi-vectoral policies – which consist in striking up relations with a range of actors – will guarantee Kazakh independence.

Chinese pursuit of Turkmen gas

Given China's commitment to tripling the share of gas in its energy mix from 4 per cent to 12 per cent by the end of the decade, Turkmenistan seems a promising partner (Batovic 2014). The take-off of Beijing's relations with Ashgabat dates back to 2006, when China officially decided to import gas from the Central Asian country (Paramonov *et al.* 2011b). Since then, a 35-year production-sharing agreement (PSA) has been signed between the two countries, allowing China to develop the Bagtyyarlyk field in Eastern Turkmenistan (Paramonov *et al.* 2011b). The agreement should enable Turkmenistan to feed gas into the Central Asia-China pipeline, currently stretching 7000 km from Turkmenistan, through Uzbekistan and Kazakhstan to China.

Other important investments include US$8.1 billion in Chinese loans for Turkmengaz with the aim of supporting the development of the country's largest gas field – Galkynysh – situated in the east of Turkmenistan (Hasanov 2009). An agreement for a new pipeline was also signed in September 2013 during the Chinese presidential visit to the Central Asian country, which would increase exports to 65 billion cubic metres a year (Platts 2013). The endeavour, which was financed by loans provided by the Development Bank of China, led to a joint declaration being signed on the strategic partnership between Turkmenistan and China, highlighting the importance of Turkmenistan in China's energy security strategy (Genté 2013).

Trade explosion with Uzbekistan

While Sino–Uzbek relations were underdeveloped in the years after independence, they began to thrive from 2007 onwards after China started granting credits to Uzbekistan for the purchase of Chinese goods (Paramonov *et al.* 2011a). At present, China is Uzbekistan's second trade partner and first foreign investor with a 35 per cent share of total investments (Najbullah and Babajanov 2013). Tashkent exports are mainly concentrated in cotton fibre, non-ferrous metals and fertilisers, as well as gas, although the volumes of the latter are inferior to those originated in Turkmenistan (Tolipov 2013).

While enjoying an increase in available gas reserves, Tashkent has steadily become more interested in diversifying its list of exporters. As most gas is currently being exported to Russia – thus according it leverage over Tashkent – alternative routes are being sought, namely to China and the Pacific Ocean (Hodzhaev 2010). In 2007, Uzbekistan decided to join the Chinese pipeline project spanning Turkmenistan, Uzbekistan and Kazakhstan, and pledged to provide 10 of the total 40 billion cubic metres (Hodzhaev 2010). In turn, Beijing made US$74 million available for the modernisation of the Uzbek gas distribution network in order to strengthen gas flows to China (US Energy Information Administration 2012a).

As the demand for gas rises in China, only Russia and the Central Asian states are deemed to be among the available overland suppliers (Hodzhaev 2010). However, as Russia needs time to build pipelines from eastern to western Siberia before it can connect to the Chinese market and there are yet-to-be-resolved price disputes between Moscow and Beijing, it appears that Turkmenistan, and to some extent Uzbekistan, will fulfil this role in the short to medium-term (Hille and Hornby 2014). This will however partly depend on the development and exploration of potential reserves through the attraction of foreign investors. Moreover, Uzbekistan's inefficient infrastructure must be modernised and more pipelines must be built to counter its landlocked position (Okumuş 2013).

As Uzbekistan currently has limited resources to export, in comparison with Turkmenistan and Kazakhstan, it is believed that Chinese interest in the country is more geopolitical and strategic rather than purely economic (Paramonov *et al.* 2011a) This stems from the fact that Uzbekistan is the infrastructural heart of Central Asia, as well as being the most populous country without which regional projects are deemed futile. Its relative urbanisation and industrialisation are further assets it can boast compared to other Central Asian countries, which also explain the large number of joint ventures between Chinese and Uzbek companies (Umarova 2013).

Transit corridor and export hub: Kyrgyzstan

Although flows have tended to be informal, trade relations have consistently deepened since 1991. Because Kyrgyz taxes are lower for imports transported on a small-scale by individuals than in other neighbouring countries, Kyrgyzstan has become a base for the re-export of Chinese goods to Russia, Kazakhstan and Uzbekistan (Mogilevskii 2012: 38). In terms of informal trade, this comprised 200 per cent of total official imports, whereas informal re-exports represented 640 per cent of total official Kyrgyz exports in 2008, illustrating the importance of such informal activity (Mogilevskii 2012: 43). Indeed, Kyrgyzstan has in less than a decade become the 'entrepôt for Chinese consumer goods in Central Asia' (Kaminski and Raballand 2009) on the way to the Middle East, Russia and South Asia, with around 800,000 traders benefiting from such re-export business (Rotar 2012). Two main reasons explain Kyrgyzstan's rise as transit corridor for Chinese goods. First, its geographic location between Xinjiang and three Central Asian republics makes it a favourable transit bridge for Chinese trade westwards. Second, as a WTO member, China prefers other members for trade business in order to benefit from lower trade barriers. In this regard, as the only Central Asian WTO member until 2013,[20] Kyrgyzstan had a comparative advantage vis-à-vis its neighbours.

China has heavily invested in Kyrgyz infrastructure, rebuilding motorways connecting the main economic centres of the country with its borders. In the

same vein, China has pushed for the development of a railway network connecting China, Kyrgyzstan, Uzbekistan, Iran and Turkey with the European
rail network, in order to expedite the shipping process from China to Europe.
Beijing has also invested in Kyrgyzstan's energy network, targeting oil refineries and the electricity network among other sectors (Shlapentokh 2013). In
doing so, it has partly broken Russia's monopoly over fuel supply, which was
previously used to apply pressure in the political field (Rickleton 2013a). The
fact that Line D would pass through Kyrgyzstan is of equally great importance, for it would enable Kyrgyzstan to earn extra revenue and increase its
leverage vis-à-vis its neighbours.

Finally, Kyrgyzstan is increasingly seen as an alternative market for raw
material imports due to its substantial uranium, gold, coal, lead, zinc and
antimony deposits (Brown 2010). While Russia can cater for Chinese needs,
Kyrgyzstan is seen as a reliable alternative and perhaps a more willing partner. China's active pursuit of raw materials in the country is reflected in the
investment of Sinosteel in the exploration of uranium, which if successful will
ensure 60 per cent ownership of mines (World Nuclear Association 2013).

Beijing has generally been open to providing Kyrgyzstan with loans with no
strings attached. The latest example is US$3 billion worth in grants and
credits during Xi's September visit to Bishkek (Syroezhkin 2013). While this
enables it to replenish its budget, it also risks leading to a loss of sovereignty
in case the government defaults (Kostenko 2013). Chinese dominance in
domestic political and economic life would further fuel anti-Chinese sentiment. This is already being stoked by perceived inequalities in the sphere of
Sino–Kyrgyz labour relations: the Chinese are criticised for hiring their own
nationals and bringing their own equipment, thus not contributing to the
development of Kyrgyzstan (Asanov 2013). In a country where thousands
leave for Russia and Kazakhstan in search for work, the presence of an estimated 50,000 Chinese workers could well be a bone of contention (Asanov
2013). At the same time, the Chinese are notorious for providing bad working
conditions, casting doubt on the real magnitude of public resentment over the
hiring of Chinese nationals.

The Tajik credit 'muncher'

Given the instability in Tajikistan during the civil war period, it is not
surprising that relations took the longest to develop between Beijing and
Dushanbe. Even if Tajikistan and China share a border, geographical as
well as infrastructural factors greatly complicate the process of bringing goods from China to Tajikistan. Despite these inconveniences, China
is Tajikistan's greatest investor and principal trade partner (Sodiqov
2012). The latter status is partly supported by the re-export of Chinese
goods, which move on from Tajikistan to Afghanistan and potentially
Uzbekistan – although to a far lesser extent than in Kyrgyzstan (Mogilevskii
2012: 38).

In terms of development, it can be seen that a similar strategy as the one currently in place in Kyrgyzstan has been adopted in Tajikistan. While China provides these countries with much-needed investments and loans, it simultaneously exploits its mineral and energy potential (Sattori 2013). Chinese investments have ranged from the mining sector to exploration for oil and gas, as well as infrastructural projects, all of which should help kick-start Tajik development (Sattori 2013).

In terms of debt, Tajikistan owes China US$862 million, an amount it may very possibly default on. While there are similar fears as in Kyrgyzstan, there is additional resentment concerning the transfer of 1,158 square kilometres from Tajikistan to China, which took place in 2011 (Vinson 2012). Furthermore, China is also accused of paying unfair wages to local workers and preferring to hire its own nationals, as is the case in Kyrgyzstan. However, given the greater poverty levels in Tajikistan and its dependence on remittances, this may be an even more bitter pill to swallow for the Tajik (Vinson 2012).

Appeasing Russia and pacifying Xinjiang: multilateralism made in Shanghai

While China has the economic power leverage in Central Asia, its position vis-à-vis Russia is much less hierarchical – if at all. From Beijing's point of view, the best way to appease Russia while keeping its economic influence in the region consists in embedding Moscow in a multilateral framework. This option is feasible, since both countries share common interests and threats in the security field. Pronouncing the word 'security' in a Chinese context, however, makes speaking about Xinjiang unavoidable. China has found a way to tackle both challenges in multilateralism and has pushed for the development of the Shanghai Cooperation Organisation (SCO). As their main multilateral cooperation instrument in Central Asia, it serves as a balance for competing interests and agendas on the one hand, and common concerns on the other, all the while helping to build trust in Central Asia.

Borders and Xinjiang at the origin of the Shanghai Cooperation Organisation

In order to understand the origins of this regional cooperation, one must go back to the late 1980s, a period that was marked by a rapprochement between the USSR and China, as well as a fervent desire to settle decades-old border disputes. The rather sudden dissolution of the Soviet Union triggered new and serious concerns in Beijing, fearing a destabilising effect in its neighbourhood. Beijing was therefore eager to discuss border and military issues with the new republics and created the so-called '4+1 talks' involving China on the one side and its direct neighbours, Russia, Kazakhstan, Kyrgyzstan and Tajikistan, on the other. These talks were held on an annual basis and eventually became the foundation of the Shanghai Five in 1996 (Wacker 2011: 70–73).

Having become official in June 2001, the SCO accepted the full membership of Uzbekistan and continued to expand with Mongolia, India, Iran, Pakistan and Afghanistan being granted observer status in the years that followed (Ambrosio 2008: 1327), and its scope was further enhanced by adding a number of dialogue partners to the list.[21] As a membership condition, the above-mentioned participants pronounced their commitment to 'mutual trust, mutual benefit, equality, consultation, respect for multi-civilisations and pursuit of common development' (SCO 2006). With time, political, economic, security and diplomatic objectives (Ambrosio 2008: 1326) were supplemented with cooperation in the fields of energy, communication, transport and agriculture (Kassenova 2010: 164)

Not long after its creation, the SCO caused nervous reactions, particularly in the West, with labels such as the 'dictators' club' (Tisdall 2006) or the 'NATO of the East' (Crisell 2008), drawing a sombre picture of the organisation. The SCO Charter of 2001 emphasises 'mutual respect of sovereignty, independence, territorial integrity of states and inviolability of state borders, non-aggression, non-interference in internal affairs, non-use of force or threat of its use in international relations' (China Daily 2006). It thus stands in contrast with the Western-dominated security vision, which advocates and actively promotes the human security and good governance dimension. With its strong focus on the combat of the 'three evils', terrorism, extremism and separatism, the SCO's security vision is reported to further reinforce authoritarian practices in the region (Ambrosio 2008: 1334; Von Hauff 2013) and to be a 'vehicle for human rights violation' (FIDH 2012).

The SCO principles prove that Russia and China have an overlapping normative basis that guides their behaviour in international relations and allows them to find common ground with the states of Central Asia. However, beyond this compliance, the SCO is presented as being China-driven. Remarkably, 'the SCO is the only multilateral grouping in which China is involved that has from the first been driven by Beijing' (Clarke 2010: 122) and that has also its headquarters in the Chinese capital. Indeed, the SCO is an initiative that best translates China's wider foreign policy paradigm: away from its traditional isolationism towards an active shaping of multilateralism in its vicinity.

Stable, friendly and institutionalised relations with its Central Asian neighbours give China the opportunity to also strengthen new markets on its western periphery, to develop new supply sources for its energy demand and, most importantly, to enhance geopolitical security in its vicinity. By focusing on separatism, terrorism and extremism, China appeals to an interest common to all SCO member states, namely the combat of internal challenges to their regime security (Aris 2009b: 462). At the same time, Beijing has found a tool to contain the Islamic separatist groups in Xinjiang, its agitated Autonomous Region in the northwest, with the support of external partners. One of the biggest concerns on Beijing's SCO agenda was therefore to convince Central Asian leaders to prevent solidarity with this movement from Central Asia's considerable Uighur diaspora.[22]

Sino–Russian diverging interests in the SCO

China's objectives in the organisation are today manifold, though overlap with its broader interests in Central Asia. In essence, the SCO can be described as its main vehicle of cooperation with Russia and a soft tool to enter Central Asia via a multilateral approach without triggering fears of a new wave of imperialism on the part of the sovereignty-cherishing Central Asian republics. Thus, in a wide charm offensive, Beijing tries to lessen Central Asian fears through exchanges and education programmes.[23] For instance, the 30,000 scholarships provided to students from SCO member countries can be interpreted as an attempt to ease anti-Chinese sentiment (Beshimov and Satke 2013).

As far as Russia is concerned, it is eager to (re-)amplify its influence in the energy realm, which it has been losing out in the Central Asian hydrocarbon market to the benefit of China. At a Shanghai summit in June 2006, Russia hence proposed the creation of an energy club that would serve as a unified market for all oil, gas and electricity exporters, consumers and transit countries while equally serving as a platform for the coordination of energy policies and investments (Overland *et al.* 2009: 162). While China is the most enthusiastic SCO member when it comes to expanding cooperation domains, it has been far less interested in the Russian energy club initiative. Since it has widely assured its gas and oil imports from the region, Beijing would not benefit from a possible cartel regulating hydrocarbon prices. Besides, energy does not seem to be China's major interest in the SCO, otherwise it would have strongly supported Iran's membership in the organisation, one of its major gas suppliers. Interestingly, the SCO Charta denies membership to any country under UN sanctions – a clause that China has been insisting on itself. The energy club therefore does not live up to its potential and is, due to Chinese hesitations, 'little more than a consultative body to discuss already existing cooperation' (Raith and Weldon 2008).

In Beijing's eyes, the SCO is still envisioned mainly as a security provider in Central Asia and its Xinjiang region. However, beyond the energy realm, Russia is not interested in intensifying cooperation within the SCO; it rather seeks to safeguard the influence it built up through the creation of its own regional tools: the CSTO in the military and security realm and the EurAsEC in the trade domain. It has therefore not been enthusiastic about China's plans to reduce trade barriers, which would further flood local markets with Chinese goods and reduce Russia's share. Given these divergent utility projections, the perspectives for the development of the SCO's economic and energy dimension require prudent assessment. It has become increasingly apparent that Russia and Central Asian countries fear economic absorption by the Chinese economic giant. They are aware of the fact that their industries are far from able to compete with Chinese labour and manufactures. It is then of no surprise that Beijing's initiative of creating a SCO Free Trade Zone was unanimously rejected by Russia and the Central Asian republics.

Similarly, China's request to join the EurAsEC met with their refusal (Kaukenov 2009: 38f.).

Indeed, China and Russia are seen as the backbone of the organisation, but the Central Asian states do have the power to veto any decision they do not deem in their interest (Aris 2008: 6). Be that as it may, the SCO has been described as asymmetrical due to its composition. With two global powers and four states that can be described as comparatively weak, there is great inequality in terms of economic, political, military and demographic strength (Tolipov 2007: 29). Nevertheless, in the eyes of Central Asian leaders, whose countries are sandwiched between the two regional hegemonies, the SCO framework can be a useful 'buffer to balance Russian and Chinese influence in the region' (Tumurkhuleg 2012: 187).

Mutual containment in Central Asia

Beijing needs Russia to be in the SCO for two main reasons. Firstly, the multi-lateral framework of the organisation keeps Sino–Russian relations stable and enhances their cooperation in the numerous common fields of interest. More importantly, however, Russia is an indispensable part of China's prudent diplomacy in Central Asia. China is aware of the neo-colonialist threat scenarios it creates in the region due to its economic and demographic weight. Therefore Russia's involvement serves as a guarantee against the appetite of a too strong neighbour.

Containment is also Russia's main interest in the SCO, even though its motivation differs. On the international level, Moscow and Beijing overtly demonstrate their union. For instance, they pursue the same agenda in the UN Security Council when it comes to condemning external interference or to preventing intervention in conflict areas such as Syria. Hence, their apparent honeymoon within the SCO triggered serious concerns of an 'Anti-Western Alignment' (e.g. Aris 2009a; Spencer 2006). However, a closer look allows another picture to emerge, which is tainted by competition over influence, mutual supervision and reluctance to develop genuine cooperation in strategic fields.

Russia clearly seeks to tame China's power and weight in the region. The considerable degree of mistrust between both great powers is best reflected by Moscow's refusal to sell sensitive technology equipment to China or to license the exploitation for certain weapon systems (Laumouline 2006: 12). While Russia wants to keep an eye on China's activities in a region that it continues to regard as its sphere of influence, it is reluctant to enhance further cooperation. A closer reading of Putin's foreign policy vision for Russia, which he outlined in a widely discussed article (Putin 2012) before the presidential elections, is in this regard revealing. Putin refers to China several times as an important partner who challenges 'decisions dictated by someone else'. However, he systematically embeds their cooperation in multi-lateral frameworks such as 'the U.N. Security Council, BRICS, the SCO, the G20 and other multilateral forums' (ibid.). The most striking observation is the fact that this is the only time Putin refers to the SCO, while the other forums enjoy more attention in the text. The SCO therefore does not have a prominent place on Russia's foreign policy agenda. However, it is unlikely that Russia will renounce this unique opportunity to watch China's movements in Central Asia, which is given by the SCO framework. Without investing too much enthusiasm, Russia will certainly keep its seat warm in the SCO.

Conclusion

This chapter has sought to illustrate China's rise in Central Asia, which is largely taking place in the field of trade and energy. While there is some variation, it is evident that China is largely pushing Russia aside in these spheres and relegating it to a lower position. This stems from the fact that Beijing needs hydrocarbon reserves and raw materials to fuel its development and the Central Asian states can meet its requirements. However, the proximity to China is at the same time an opportunity and a source of worry for Central Asia. Beijing therefore permanently has to perform a fragile balancing act of assuring its status as an economic force without fuelling neo-colonialist fears in the Central Asian republics.

As Russia no longer has the leverage to make Central Asia its sole back-yard and is forced to cohabit with China, the Shanghai Cooperation Organisation should increasingly be employed as a far-reaching instrument for the coordination of Sino–Russian interests. This statement is all the more true in light of President Xi's extensive 2013 visit, which has further intensified the Chinese presence in the economic and energy sector. Moreover, as Sino-Russian relations deepen outside of the region in the context of new oil deals between Moscow and Beijing, the dynamics within the region are likely to change simultaneously. This is because Moscow will not be able to ignore its partner's interests in Central Asia once its own budget will depend on Chinese purchases of oil and gas.

Beyond the function of a coordinator of Beijing and Moscow's interests, the multilateral approach via the Shanghai Cooperation Organisation is a tranquiliser in Chinese hands to ease fears of neo-imperialism towards its Central Asian partners. The cooperation with Russia within the SCO therefore serves as a guarantee of balance in the region and as a key to Central Asian markets and raw materials.

Despite Chinese affirmations that its rise will be peaceful and it has no political aspirations, the room for manoeuvre of all states concerned will be reduced according to the growth of its neighbour's economy and political status in the global arena. However, as none of the leaders of the region are oblivious to the risks of Chinese expansion, other actors such as the USA, the EU, India, Turkey and Russia will remain close partners of the Central Asian states in an attempt to safeguard their sovereignty through multi-vectorism.

Notes

1 Near abroad is a geopolitical term used to refer to the former Soviet Republics.
2 Besides oil and gas, the countries of the region also have substantial coal, iron, uranium and nonferrous metal reserves.
3 The Commonwealth of Independent States (CIS) combined – the successor organisation of the Soviet Union – only presented 3.8 per cent of total trade with China in 2012, of which 2.4 per cent was with Russia alone. See European Commission 2012.
4 According to the Logistics Performance Index of 2007 and 2010, every single Central Asian country improved in the ranking, although greatest progress could be observed in Uzbekistan and Kazakhstan.
5 The Kazakhstan-China oil pipeline stretches 2558 km from Atyrau in Kazakhstan to Xinjiang in China. It was finalised in 2009 as a joint venture between CNPC and KMG. See US Energy Information Administration 2013.
6 Examples of these are Godemont 2011, and Cooley 2012.
7 President Karimov's speeches emphasising Uzbek sovereignty from Russia are indicative of this.
8 Created in 2002 and comprising the former Soviet republics (Armenia, Belarus, Kazakhstan, Kyrgyzstan, Russia and Tajikistan; Uzbekistan left the organisation in 2012), the CSTO is the main regional institution with a military dimension.
9 EurAsEC aims at establishing a Single Economic Space and at strengthening economic and trade ties between the member states (Russia, Belarus, Kazakhstan, Kyrgyzstan and Tajikistan; Uzbekistan suspended its membership in 2008).
10 A single customs tariff began to be applied in Russia, Kazakhstan and Belarus on 1 January 2010, which also resulted in Russian WTO commitments becoming binding for the other two states. This Customs Union was then supposed to gradually develop into the Eurasian Economic Union.
11 According to statistics, the turnover of goods between Russia and Central Asia is to have dropped ten times between 1991 and 1992, only recovering from 2003 onwards. Kazakhstan and Uzbekistan currently carry out the largest amount of trade with Russia as they provide it with hydrocarbons. As Turkmenistan reduced its gas transfers to Russia, it renounced its previous position of pre-eminence and moved closer in the direction of Tajikistan and Kyrgyzstan in terms of trade flows.
12 The Russian fear of Chinese monopolisation of its hydrocarbon exports expressed itself in Moscow's decision to reject the possibility of building an oil pipeline from East Siberia to Daqing in Manchuria in 2004. Instead, a diversification strategy was adopted, with the pipeline being extended to Perevoznaya despite the substantial costs the option entailed. It did however allow Moscow to entertain the idea of exporting to the wider Asia Pacific region and thus expanding the list of its customers.
13 15 million tonnes of oil are currently sent to China a year by means of the East Siberia Pacific Ocean pipeline.
14 In October 2012, several hundred inhabitants of the village Orlovka (Northern Kyrgyzstan) protested in front of the Chinese company Zijing Mining Group and threatened to destroy the offices. As a consequence, 250 Chinese employees were evacuated. One year before, in August 2011, three Chinese miners were beaten up by a mob of 300 locals in the Naryn Province of Kyrgyzstan. See Rotar 2012.
15 In the 1960s and 1970s, production lay around 500,000 barrel per day, and it began to rise in the mid-1990s (530,000 barrel per day in 1992), exceeding 1 million barrel per day in 2003. Taken from U.S. Energy Information Administration 2013, and Caspian Mainport 2009.
16 Chevron-Texaco, Exxon Mobil, Shell, TotalFinaElf, British Gas, Statoil, Eni-Agip and Philips Petroleum were among the main investors.

17 Dutch disease is defined as 'the negative impact on an economy of anything that gives rise to a sharp inflow of foreign currency, such as the discovery of large oil reserves. The currency inflows lead to currency appreciation, making the country's other products less price competitive on the export market. It also leads to higher levels of cheap imports and can lead to deindustrialisation, as industries other than resource exploitation are moved to cheaper locations'. See Financial Times Lexicon.

18 In this case, Kazakhstan gains joint control over a pipeline going through Russia.

19 Due to harsh climate conditions, investors have struggled to make profit on the field, as 5 times the original amount had to be invested even though oil has hardly flowed. However, it is the fifth largest field in the world in terms of reserves. See Daly (2014).

20 The Kyrgyz Republic was the only WTO member in the region until the accession of Tajikistan in March 2013.

21 ASEAN, CIS, Belarus, Sri Lanka and Turkey.

22 About 350 000 Uighurs live in Central Asian republics (Marketos 2009: 12), with the majority living in Kazakhstan, which is home to more than 220 000 Uighurs (Tumurkhuleg 2012: 184).

23 In addition, Beijing offers study tours for 10 000 students and teachers at the Confucius Institutes in the Central Asian republics. See Olcott (2013).

Bibliography

Ambrosio, T. (2008) 'Catching the 'Shanghai Spirit': How the Shanghai Cooperation Organization Promotes Authoritarian Norms in Central Asia.' *Europe-Asia Studies* 60(8): 1321–1344.

Aris, S. (2008) 'Russian-Chinese Relations Through the Lens of the SCO.' *Russie. Nei. Visions* No. 34, Ifri, Paris. Available at: www.ifri.org/files/Russie/Ifri_RNV_Aris_SCO_Eng.pdf (accessed 15 November 2013).

Aris, S. (2009a) 'Shanghai Cooperation Organisation: An Anti-Western Alignment?' *CSS Analyses in Security Policy* No. 66, December, ETH Zurich.

Aris, S. (2009b) 'The Shanghai Cooperation Organisation: Tackling the Three Evils. A Regional Response to Non-Traditional Security Challenges or an Anti-Western Bloc?' *Europe-Asia Studies* 61(3): 457–482.

Asanov, B. and Najibullah, F. (2013) 'Kyrgyz Ask Why Jobs At Home Are Going to Chinese.' *Radio Free Europe* (6 November). Available at: www.rferl.org/content/kyr gyzstan-chinese-jobs-unemployment/25170163.html (accessed 18 November 2013).

Balci, B. (2014) 'In Taking Crimea, Putin Will Lose Central Asia and the Caucasus.' *Foreign Policy Journal* (24 March). Available at: www.foreignpolicyjournal.com/2014/03/24/in-taking-crimea-putin-will-lose-central-asia-and-the-caucasus/ (accessed 12 April 2014).

Batovic, A. (2014) 'Russia is Eyeing China to Expand Energy Ties With Asia.' *Global Risk Insights* (24 February). Available at: http://globalriskinsights.com/2014/02/24/russia-is-eying-china-to-expand-energy-ties-with-asia/ (accessed 12 April 2014).

BBC (2014) 'Kyrgyzstan's Dilemma Over Russian-Led Customs Union', *BBC*, 20 January. Available at: www.bbc.com/news/world-asia-25718770 (accessed 12 April 2014).

Beshimov, B. and Satke, R. (2013) 'China Extends Grip in Central Asia', *Asia Times*, 13 November. Available at: www.atimes.com/atimes/Central_Asia/CEN-01-131113.html?goback=%2Egmr_4208659%2Eamf_4208659_40000925%2Egde_4208659_mem ber_5808506169391591425#%21 (accessed 18 November 2013).

Bierman, S. and Arkhipo, I. (2013) 'CNPC Buys Stake in Novatek's Yamal LNG Project in Russian Arctic', *Bloomberg*, 5 September. Available at: www.bloomberg. com/news/2013-09-05/cnpc-buys-stake-in-novatek-s-yamal-lng-project-in-russian-arc tic.html (accessed 17 September 2013).

Bierman, S. (2013) 'Rosneft Agrees to Sell Sinopec $85 Billion of Oil Over 10 Years', *Bloomberg*, 22 October. Available at: www.bloomberg.com/news/2013-10-22/rosneft-a grees-to-sell-sinopec-85-billion-of-oil-over-10-years.html (accessed 18 December 2013).

Brown, D. (2010) 'Uranium Mining in Kyrgyzstan', *Uranium Investing News*, 21 September. Available at: http://uraniuminvestingnews.com/4818/uranium-mini ng-in-kyrgyzstan.html (accessed 19 November 2013).

Burashev, M. (2009) 'The Sale of MangistauMunaiGas', *Halyk Finance*, 30 November. Available at: www.halykfinance.kz/en/site/index/research/news:79899 (accessed 18 November 2013).

Castillo, A. (2008) 'SCO: Rise of NATO East?' *Diplomatic Courier, The International Relations and Security Network* (18 August). ETH Zurich. Available at: www.isn. ethz.ch/Digital-Library/Articles/Detail/?lng=en&id=90108 (accessed 15 November 2013).

Caspian Mainport (2009) Available at: www.caspianmainport.com/public/Kaz.jsp (accessed 15 November 2013).

China Daily (2006) 'Shanghai Cooperation Organisation Charter', *China Daily*, 12 June. Available at: www.chinadaily.com.cn/china/2006-06/12/content_6020341.htm (accessed 18 November 2013).

Clarke, M. (2010) 'China and the Shanghai Cooperation Organisation: The Dynamics of "New Regionalism", "Vassalisation", and Geopolitics in Central Asia.' In E. Kavalski (ed.), *The New Central Asia: The Regional Impact of International Actors.* Singapore: World Scientific Publishing.

Cooley, A. (2012) *Great Games, Local Rules: The New Great Power Contest in Central Asia.* New York: Oxford University Press.

Crisell, P. (s.d.) 'The Shanghai Cooperation Organisation: The NATO of the East?' *Newnations Special Report.* Available at: www.newnations.com/specialreports/sco. html (accessed 15 November 2013).

Daly, J. (2014) 'Kashagan, Down But Not Out.' *Oil Price* (8 January). Available at: http://oilprice.com/Energy/Crude-Oil/Kashagan-Down-but-not-Out.html (accessed 7 March 2014).

Deming, L. (2012) 'Kyrgyzstan Still a Mine Field for Investors', *Global Times*, 28 October. Available at: www.globaltimes.cn/DesktopModules/DnnForge%20-% 20NewsArticles/Print.aspx?tabid=99&tabmoduleid=94&articleId=740848&moduleI d=405&PortalID=0 (accessed 18 December 2013).

Demytrie, R. (2010) 'Struggle for Central Asian Energy Riches', *BBC*, 3 May. Available at: www.bbc.co.uk/news/10175847 (accessed 28 November 2013).

Dyomkin, D. (2013) 'Russia Grabs China Oil and Gas Export Deals', *Reuters*, 22 October. Available at: www.reuters.com/article/2013/10/22/china-russia-ener gy-idUSL5N0IC10F20131022 (accessed 18 October 2013).

EBRD Transition Report (2013) 'Country Assessments Uzbekistan'. Available at: http://tr.ebrd.com/tr13/en/country-assessments/3/uzbekistan (accessed 15 February 2014).

Economist (2013) 'China Rising, Sinking Russia', *The Economist*, 14 September. Available at: www.economist.com/news/asia/21586304-vast-region-chinas-econom ic-clout-more-match-russias-rising-china-sinking (accessed 17 November 2013).

European Commission (2012) 'European Union, Trade in goods With China.' Available at: http://trade.ec.europa.eu/doclib/docs/2006/september/tradoc_113366.pdf (accessed 13 November 2013).

Federation for Human Rights (FIDH) (2012) 'Shanghai Cooperation Organisation: A Vehicle for Human Rights Violations.' *FIDH* (18 October). Available at: www.fidh. org/en/eastern-europe-central-asia/Publication-of-a-report-Shanghai-12031 (accessed 15 September 2013).

Genté, R. (2013) 'China Takes a Decisive Advantage in Turkmenistan.' *Turkmen Chronicles* (27 September). Available at: www.chrono-tm.org/en/2013/09/china-ta kes-a-decisive-advantage-in-turkmenistan/ (accessed 5 November 2013).

Goble, P. (2013) 'Uzbekistan Faces Civil War, Possible Disintegration, Tashkent Scholar Says.' *Eurasia Daily Monitor* 10(164) (17 September), Jamestown Foundation. Available at: www.jamestown.org/single/?tx_ttnews[tt_news]=41369&no_ca che=1#.VK3D4CeN-YU (accessed 18 November 2013).

Godement, F. (2011) 'The New Great Game in Central Asia.' *European Council of Foreign Relations.* Available at: www.ecfr.eu/page/-/China%20Analysis_The%20new %20Great%20Game%20in%20Central%20Asia_September2011.pdf (accessed 25 September 2013).

Hale, H. E. (2013) 'Age is No Friend to Dictators.' *PONARS Eurasia* (27 March). Available at: www.ponarseurasia.org/ru/node/6119 (accessed 15 November 2013).

Hasanov, G. (2009) 'China Granted Turkmengaz State Concern Loan of Many Billions.' *Trend* (25 June). Available at: http://en.trend.az/capital/energy/1493827. html (accessed 18 November 2013).

Hayward, D. (2009) 'China's Oil Supply Dependence.' *Journal of Energy Security* (18 June). Available at: www.ensec.org/index.php?option=com_content&id=197:china s-oil-supply-dependence&catid=96:content&Itemid=345 (accessed 15 October 2013).

Hille, K. and Hornby, L. (2014) 'Gazprom Close to Agreeing Pricing Deal on China Gas Supplies', *Financial Times*, 5 January. Available at: www.ft.com/intl/cms/s/0/ 38b246ba-6bb9-11e3-85b1-00144feabdc0.html#axzz2w43S0akw (accessed 5 March 2014).

Hodzhaev, A. (2010) 'Gazoprovod "Tsentral'naya Aziya – Kitay": Diversifikatsiya Eksportnykh Potokov Gaza [Central Asia-China Pipeline: Diversification of Gas Export Flows].' *Centre for Political Studies.* Available at: http://cps.uz/ru/analitika -i-publikatsii/gazoprovod-%c2%abtsentralnaya-aziya-%e2%80%93-kitai%c2%bb-div ersifikatsiya-eksportnykh-potokov (accessed 15 December 2013).

Hydrocarbons Technology 'Kazakhstan-China Crude Oil Pipeline, Kazakhstan.' Available at: www.hydrocarbons-technology.com/projects/kazakhstan-china-cru de-oil-pipeline/ (accessed 18 November 2013).

Indeo, F. (2013) 'The Impact of the "Shale Gas Revolution" on Russian Energy Strategy.' *EGS* Working Paper 2013–2018, Center for Energy Governance & Security. Available at: http://nautilus.org/napsnet/napsnet-special-reports/the-impa ct-of-the-shale-gas-revolution-on-russian-energy-strategy/#ixzz2w3BuO3d3 (accessed 15 December 2013).

International Crisis Group (2011) 'Tajikistan: The Changing Insurgent Threats.' *Asia Report* No. 205. Bishkek/Brussels, 24 May 2011.

International Crisis Group (2012) 'Kyrgyzstan: Widening Ethnic Divisions in the South.' *Asia Report* No. 222. Bishkek and Brussels, 29 March.

Jarosiewicz, A. (2013) 'A Chinese Tour de Force in Central Asia.' *Centre for Eastern Studies* (18 September). Available at: www.osw.waw.pl/en/publikacje/ea stweek/2013-09-18/a-chinese-tour-de-force-central-asia (accessed 5 September 2013).

Jonson, L. (1998) *Russia and Central Asia: A New Web of Relations, Central Asian and Caucasian Prospects.* London: The Royal Institute of International Affairs.

Kalicki, J. (2001) 'Caspian Energy at a Crossroads.' *Foreign Affairs* (September/ October). Available at: www.foreignaffairs.com/articles/57244/jan-h-kalicki/caspia n-energy-at-the-crossroads (accessed 18 November 2013).

Kaminski, B. and Raballand, G. (2009) 'Entrepôt for Chinese Consumer Goods in Central Asia: The Puzzle of Re-Exports Through Kyrgyz Bazaars.' *Eurasian Geography and Economics* 50(5): 581–590.

Karimov, O. (2012) 'Russia and Uzbekistan: Oil and Gas Cooperation.' *Ria Novosti* (21 January). Available at: http://en.ria.ru/international_affairs/20100720/159879904. html (accessed 1 November 2013).

Kassenova, N. (2010) 'The Shanghai Cooperation Energy Club Purpose and Prospects.' In I. Overland, H. Kjaernet and A. Kendall-Taylor (eds), *Caspian Energy Politics: Caspian Energy Politics Azerbaijan, Kazakhstan and Turkmenistan.* London: Routledge, pp. 162–177.

Kaukenov, A. (2009) 'Chinese Diplomacy in Central Asia: A Critical Assessment.' In M. Esteban and N. De Pedro (eds), *Great Powers and Regional Integration in Central Asia: A Local Perspective.* Madrid: Exilibris, pp. 35–52.

Kogtev, Y. (2013) 'Latest Oil Deal Making Russia Dependent on China.' *Russia Beyond the Headlines* (1 September). Available at: http://rbth.asia/world/2013/09/01/ latest_oil_deal_making_russia_dependent_on_china_48667.html (accessed 1 November 2013).

Kostenko, J. (2013) 'Kyrgyzstan and China: From Strategic Partnership to Expansion', *News Agency 24*, 12 September. Available at: http://eng.24.kg/politic/2013/09/12/ 27909.html?print=yes (accessed 1 November 2013).

Laumouline, M. (2006) 'L'Organisation de Coopération de Shanghai Vue d'Astana: Un "Coup de Bluff Géopolitique"?' *Russie. Nei.Visions* No. 12, Ifri, Paris. Available at: www.ifri.org/sites/default/files/atoms/files/laumullinfrancais.pdf (accessed 15 November 2013).

Lelyveld, M. (2014) 'China Pursues New Central Asian Gas Route.' *Radio Free Asia* (10 February). Available at: www.rfa.org/english/commentaries/energy_watch/ga s-02102014124143.html (accessed 18 March 2014).

Malhotra, A. (2012) 'Chinese Inroads Into Central Asia: Focus on Oil and Gas.' *Journal of Energy Security.* Available at: www.ensec.org/index.php?option=com_ content&view=article&id=387:chinese-inroads-into-central-asia-focus-on-oil-and-ga s&catid=130:issue-content&Itemid=405 (accessed 15 November 2013).

Mariani, B. (2013) 'China's Role and Interests in Central Asia.' *Saferworld.* Available at: www.isn.ethz.ch/Digital-Library/Publications/Detail/?lng=en&id=172938 (accessed 15 July 2014)

Marketos, T. N. (2009) *China's Energy Geopolitics. The Shanghai Cooperation Organization and Central Asia. Contemporary China Series,* New York: Routledge.

Mashrab, F. (2013) 'Xi Jinping Brings Out Central Asia Critics', *Times of Central Asia,* 24 September. Available at: www.atimes.com/atimes/Central_Asia/ CEN-01-240913.html (accessed 18 November 2013).

Mason, E. (2013) 'Kazakhstan to Expand Oil Pipeline to China.' *Oil and Gas Technology* (19 September). Available at: www.oilandgastechnology.net/pipeline-news/kazakhstan-expand-oil-pipeline-china (accessed 18 November 2013).

Mogilevskii, R. (2012) 'Trends and Patterns in Foreign Trade of Central Asian Countries.' *Institute of Public Policy and Administration University of Central Asia* Working Paper No. 1. Available at: www.ucentralasia.org/downloads/UCA-Trends%26PatternsForeignTradeCA-Eng-May2012.pdf (accessed 5 September 2013).

Najibullah, F. and Babajanov, K. (2013) 'Uzbeks Home in on Chinese for Opportunity.' *Radio Free Europe* (14 September). Available at: www.rferl.org/content/uzbekistan-chinese-opportunity/25104231.html (accessed 27 November 2014).

Ng, T. (2014) 'Crime Leads Chinese Expats in Central Asia to Weigh Returning Home', *South China Morning Post*, 17 January. Available at: www.scmp.com/print/news/china/article/1407495/crime-leads-chinese-expats-central-asia-weigh-returning-home (accessed 18 June 2014).

Okumuş, O. (2013) 'Uzbekistan's Strategy to Become a Gas Exporting Country.' *Natural Gas Europe* (4 June). Available at: www.naturalgaseurope.com/uzbekistans-strategy-to-become-a-gas-exporting-country (accessed 18 September 2013).

Olcott, M. B. (2013) 'China's Unmatched Influence in Central Asia.' *Carnegie Endowment for International Peace* (18 September). Available at: http://carnegieendowment.org/2013/09/18/china-s-unmatched-influence-in-central-asia/gnky (accessed 11 November 2013).

Overland, I., Kjaernet, H. and Kendall-Taylor, A. (eds) (2009) *Caspian Energy Politics: Caspian Energy Politics Azerbaijan, Kazakhstan and Turkmenistan*. London: Routledge.

Paramonov, V., Rashidov, O., Strokov, A., Stolpovsky, O. and Sattarov, Sh. (2011a) 'Economic Presence of China in Uzbekistan.' *Central Eurasia* (3 February). Available at: http://ceasia.ru/english/economic-presence-of-china-in-uzbekistan.html (accessed 18 November 2013).

Paramonov, V., Rashidov, O., Strokov, A., Stolpovsky, O. and Sattarov, Sh. (2011b) 'Economic Presence of China in Turkmenistan.' *Central Eurasia* (22 March). Available at: http://ceasia.ru/english/economic-presence-of-china-in-turkmenistan.html (accessed 22 November 2013).

Paramonov, V., Rashidov, O., Strokov, A., Stolpovsky, O. and Sattarov, Sh. (2011c) 'Economic Presence of China in Kazakhstan.' *Central Eurasia* (24 March). Available at: www.ceasia.ru/english/economic-presence-of-china-in-kazakhstan.html (accessed 18 November 2013).

Paxton, A. (2011) 'Kazakh Opposition Calls for Halt to China Expansion', *Reuters*, 28 May. Available at: http://uk.reuters.com/article/2011/05/28/kazakhstan-china-protest-idUKLDE74R02M20110528 (accessed 19 December 2013).

Pizzi, M. (2014) 'Russia, China Sign Deal to Bypass U.S. Dollar', *Al Jazeera America*, 20 May. Available at: http://america.aljazeera.com/articles/2014/5/20/russia-china-bankdeal.html (accessed 18 June 2014)

Peyrouse, S. (2012) 'Power Differential and Security Issues in Central Asia: Threat Perceptions of China.' In R.E. Bedeski and N. Swanström (eds), *Eurasia's Ascent in Energy Geopolitics. Rivalry or Partnership for China, Russia and Central Asia? Contemporary Asia Series*. New York: Routledge, pp. 92–107.

Platts (2013) 'Ashgabat, Beijing Sign Deals to Expand Turkmen Gas Exports to China', *Platts*, 4 September. Available at: www.platts.com/latest-news/natura

l-gas/moscow/ashgabat-beijing-sign-deals-to-expand-turkmen-27370637 (accessed 18 December 2013).

Putin, V. (2012) 'Russia and the Changing World', *Ria Novosti*, 27 February. Available at: http://en.ria.ru/analysis/20120227/171547818.html (accessed 18 November 2013).

Raith, M. and Weldon, P. (2008) 'Energy Cooperation and the Shanghai Cooperation Organisation. Much Ado About Nothing?' *Eurasianet* (24 April). Available at: www.eurasianet.org/departments/insight/articles/eav042508b.shtml (accessed 18 November 2013).

Rickleton, C. (2013a) 'Kyrgyzstan: China Muscles Into Energy Market, Fueling Suspicion.' *Eurasianet* (20 March). Available at: www.eurasianet.org/node/66716 (accessed 28 October 2013).

Rickleton, C. (2013b) 'Can China's Financial Might Force Stability in Central Asia?' *Oil Price* (24 October). Available at: http://oilprice.com/Geopolitics/Asia/Can-China s-Financial-Might-Force-Stability-in-Central-Asia.html (accessed 15 September 2013).

Rotar, I. (2012) 'Chinese "Expansion" in Kyrgyzstan: Myth or Reality?' *Eurasia Daily Monitor* 9(204) (7 November), Jamestown Foundation. Available at: www.jam estown.org/programs/edm/single/?tx_ttnews[tt_news]=40077&tx_ttnews[backPid27& cHash=c29e0d38eec29b68fbbc53eb0406659e#.VK3IuCeN-YU (accessed 28 November 2013).

Rotar, I. (2013) 'Uzbekistan and Kyrgyzstan Heighten Tensions in Violent Local Border Dispute.' *Eurasia Daily Monitor* 10(17) (30 January), Jamestown Foundation. Available at: www.jamestown.org/programs/edm/single/?tx_ttnews[tt_news] =40390&cHash=268108bc1c955041c74ec07ac2f6d0b4#.VK3I9CeN-YU (accessed 28 October 2013).

Rotar, I. (2014) 'Conflicts Between Tajikistan and Kyrgyzstan Potentially Undermine CSTO and Custom Union in Central Asia.' *Eurasian Daily Monitor* 11(27) (11 February), Jamestown Foundation. Available at http://www.jamestown.org/p rograms/edm/single/?tx_ttnews[tt_news]=41953&cHash=b8dad3290a574f59ff30bad9 101d5d3e#.VK3JJieN-YU (accessed 28 March 2014).

Sattori, A. (2013) 'Tajikistan Piles on China Debt', *Asia Times*, 12 June. Available at: www.atimes.com/atimes/Central_Asia/CEN-01-120613.html (accessed 28 November 2013).

SCO (2006) 'Declaration on the Fifth Anniversary of the Shanghai Cooperation Organisation.' Shanghai. Available at: http://china.org.cn/english/features/meeting/ 171589.htm (accessed 7 January 2015).

Sadykov, M. (2013) 'Uzbekistan Expresses Guarded Interest in Russia-Led Customs Union.' *Eurasianet* (13 November). Available at: www.eurasianet.org/node/67757 (accessed 13 November 2013).

Shlapentokh, D. (2013) 'Kyrgyzstan Between China And Russia.' *Central Asia Caucasus Analyst* (20 April). Available at: www.cacianalyst.org/publications/ana lytical-articles/item/12703-kyrgystan-between-china-and-russia.html?tmpl=compone nt&print=1 (accessed 6 November 2013).

Sinitsina, I. (2012) 'Economic Cooperation Between Russia and Central Asian Countries: Trends and Outlook.' *Institute of Public Policy and Administration University of Central Asia* Working Paper No. 5. Available at: www.ucentralasia. org/downloads/UCA-IPPA-WP5-RussiaInfluence-Eng.pdf (accessed 5 February 2014).

Sodiqov, A. (2012) 'Tajikistan Attracts More Chinese Funds', *Asia Times*, 19 June.

Spaiser, O. A. (2014) 'Les Républiques d'Asie Centrale Vont-Elles Être une Nouvelle Crimée?' *Le Monde. Idées*, 15 April. Available at: www.lemonde.fr/idees/article/2014/04/15/les-republiques-d-asie-centrale-vont-elles-etre-une-nouvelle-crimee_44013 96_3232.html (accessed 15 July 2014).

Spencer, R. (2006) 'Iran's Place at Summit Raises Fears of Anti-West Alliance', *The Telegraph*, 15 June. Available at: www.telegraph.co.uk/news/worldnews/middleeast/iran/1521377/Irans-place-at-summit-raises-fears-of-anti-West-alliance.html (accessed 25 November 2013).

Syroezhkin, K. (2013) 'Kitogam vizita Si Czin'pina v Central'nuju Aziju: Kyrgyzstan [The Outcome of the Visit of Xi Jingpin to Central Asia: Kyrgyzstan].' *Carnegie Endowment* (23 August). Available at: http://carnegieendowment.org/2013/09/23/к-итёгам-визита-си-цзиньпина-в-центральную-азию-кыргызстан/gnws (accessed 15 November 2013).

Taborda, J. (2014) 'China GDP Annual Growth Rate.' *Trading Economics* (20 January). Available at: www.tradingeconomics.com/china/gdp-growth-annual (accessed 19 July 2014).

The Business Year (2012) 'From the Black Lagoon.' Available at: www.thebusinessyear.com/publication/article/6/1175/kazakhstan-2012/from-the-black-lagoon (accessed 15 July 2014).

Tisdall, S. (2006) 'Irresistible Rise of the Dictators' Club', *The Guardian*, 6 June. Available at: www.theguardian.com/commentisfree/2006/jun/06/world.comment (accessed 14 July 2014).

Tolipov, F. (2007) 'The Foreign Policy Orientations of Central Asian States: Positive and Negative Diversification.' *Acta Slavica Iaponica* 16: 23–40. Available at: www.isn.ethz.ch/Digital-Library/Publications/Detail/?ots591=0c54e3b3-1e9c-be1e-2c24-a6 a8c7060233&lng=en&id=58351 (accessed 19 July 2014).

Tolipov, F. (2013) 'What Does it Mean for Uzbekistan and China to be Strategic Partners?' *The Central Asia-Caucasus Analyst* (13 November). Available at: www.cacianalyst.org/publications/analytical-articles/item/12858-what-does-it-means-for-uzbek istan-and-china-to-be-strategic-partners?.html?tmpl=component&print=1 (accessed 15 July 2014).

Tumurkhuleg, T. (2012) 'Does the Shanghai Cooperation Organisation Represent an Example of a Military Alliance?' In R. E. Bedeski, and N. Swanström (eds), *Eurasia's Ascent in Energy Geopolitics. Rivalry or Partnership for China, Russia and Central Asia? Contemporary Asia Series.* New York: Routledge, pp. 179–198.

Umarova, I. (2013). 'Uzbek, Chinese Entrepreneurs Meet in Business Forum', *Uzbek National News Agency*, 28 November. Available at: http://uza.uz/en/politics/3862/ (accessed 15 July 2014).

US Energy Information Administration (2012a) 'Uzbekistan.' Available at: www.eia.gov/countries/cab.cfm?fips=UZ (accessed 17 July 2014).

US Energy Information Administration (2012b) 'Turkmenistan.' Available at: www.eia.gov/countries/cab.cfm?fips=TX (accessed 15 July 2014).

US Energy Information Administration (2013) 'Kazakhstan.' Available at: www.eia.gov/countries/cab.cfm?fips=KZ (accessed 14 July 2014).

Vinson, M. (2012) 'China Land Deal Upsets Tajiks', *Asia Times*, 17 February. Available at: http://atimes.com/atimes/Central_Asia/NB17Ag01.html (accessed 15 July 2014).

Von Hauff, L. (2013) 'A Stabilizing Neighbor? The Impact of China's Engagement in Central Asia on Regional Security.' *DGAP Analyse* 3 (April).

Wacker, G. (2011) 'China and Its Central Asian Neighbours.' In Y. Hao and B. K. P. Chou (eds), *China's Policies on its Borderlands and the International Implications.* Macau: University of Macau, pp. 69–91.

World Nuclear Association (2013) 'Uranium in Kyrgyzstan.' Available at: http://world-nuclear.org/info/Country-Profiles/Countries-G-N/Kyrgyzstan/ (accessed 15 July 2014).

Yep, E. (2013) 'The Russian-China Oil-Export Equation.' *Wall Street Journal* (28 October). Available at: http://blogs.wsj.com/moneybeat/2013/10/28/the-russia-china-oil-export-equation/ (accessed 10 July 2014).

8 China's energy security and Sino–African energy cooperation

Ka-ho Yu and Zhou Yunheng

Introduction

This chapter examines the rationale behind China's energy investments in Africa and the mechanism of how these overseas activities could contribute to the country's energy security. We begin with China's energy diplomacy as part of the country's going-out strategy, which underpins Sino–African energy cooperation. This is followed by a discussion as to whether this energy diplomacy is merely a strategy of energy competition. The analysis then extends to the cooperation between the two powers in different energy sectors of Africa. These issues lead to a discussion about China's advantages and concerns in investing in Africa. The chapter concludes by exploring whether China's investment in Africa could encourage global energy governance in the region.

Since China will be one of the key energy consumers and producers in the next few decades, the security of energy supply has become an increasingly important factor for its domestic and international ambitions. Domestically, in order to maintain its authority, China needs to meet the economic and nationalistic expectations of its people. As Breslin (2005: 349) argues, 'it is an unwritten social contract between the party and the people whereby the people do not compete with the party for political power as long as the party looks after their economic fortunes'. Since economic development implies a greater need for energy, a stable and sufficient supply of energy is crucial to the national values and objectives of the PRC's governance (Xu 2007). Internationally, a country of the size of China is part of both the source and the solution to all major international major problems. Striving to rise as a responsible and imposing power, China cannot ignore its potential impact on the world scene. At the present, China relies heavily on conventional resources such as, oil, gas and, most importantly, coal, for power generation. In order to enhance its energy security, China is adopting the going-out strategy to diversify its energy source, and Africa, rich in resource, is on China's radar. As a latecomer to Africa, while China has its own advantages in establishing an energy relationship with African countries, its Chinese presence does not only represent new competition, but also aggravated local grievance. More importantly, China's energy strategy projects strong Chinese characteristics, which contradict Western norms. There has been a long debate concerning whether China's energy security is peaceful or militant in the international community.

China's energy security: peaceful or militant?

China's energy strategy in Africa, which has obvious diplomatic characteristics, is reducing its dependence on a sole energy supply region and sole energy transport route. Through bilateral and international energy cooperation, China is increasing its oil reserves and diversifying its supply sources, and a Chinese energy supply system is gradually being built. Since the sustainable development of the Chinese economy is closely linked to the sustainable development of the world economy, the world has started to keep an eye on China's moves.

Although the former Chinese president Hu Jintao attempted to project 'peaceful development', the outgoing strategy had the opposite effect. Such practice might exacerbate global concerns about China's actual intentions as it appears to contradict Hu's 'peaceful development'. Mearsheimer (2006, 2010) and Walt (2011b) believe that China will have an unpeaceful rise because its growth in power will lead to intense security competition or potentially even to war. China is perceived as adopting an aggressive energy diplomacy because it purchases energy resources through bilateral deals from Africa and South America instead of buying energy from the open market (Baghat 2006; Yergin 2006; Chen and Jaffe 2007). Since energy 'could be a catalyst for conflict' (Calder 2006), it could lead to resource competition and even to military races. This will increase anxiety in East Asia about safe access to overseas resources and sea lanes for transporting resources. For instance, Lim (2005: 141) claims 'China's growing strategic pressure is beginning to cause Japan to become more anxious about its security'. Despite the fact that developing countries may see China's energy hunt as a boon, other powers such as the USA and Japan are concerned about the instability and insecurity it may lead to (Zweig 2005; Zweig 2009; Zweig 2010). The USA purports in a national report that, while China claims to be keeping to its peaceful path, it is acting as if it can somehow 'lock up' energy supplies or seek direct markets and support resource-rich countries without regard to their misbehaviour (The White House 2006). Although some literature argues that China's intention is always mistakenly assumed to be locking up resources or harming the international energy supply (Garrison 2009), it is generally understood that energy import-dependent states tend to expand their influence in the global market by increasing relative and absolute power, and China is no exception (see Bajpaee 2005; Karon 2006; Mufson 2006).

Under the Chinese government's encouragement, Chinese national energy companies have been securing exploration and supply agreements with regions rich in energy (Zweig and Jianbai 2005). Moreover, China considers Africa to be a very supportive partner after it was in dispute with the West, especially during the post-Tiananmen Square period (Taylor 2007). On the other hand, according to Taylor, African countries are seeking an economic partner that, unlike the West, does not question democracy and human rights or impose political conditions. China takes the advantage of anti-hegemonies to promote its interest and position in Africa (Taylor 2006). Although the Sino–African relationship has been built on 'non-interference' in domestic affairs, the more dependent China is on African energy resource, the more likely it is to be bound to African problems. Ross (2008) argues that oil wealth can trigger conflict since it leads to economic instability, insurgencies and separatism. African countries like Angola, Congo, Equatorial Guinea, Nigeria and Sudan are suffering from these problems and China cannot neglect the impact it has, as well as its responsibility in the matter. For example, under the expression 'non-interference', one danger that China is facing in its relations with Angola is that the country allows the elites in Luanda to continue to be corrupt. Ignoring governing norms will eventually result in an unstable investment environment for Chinese companies (Taylor 2007). Another example where China's investment in Africa has attracted controversy is Sudan. China has been keen to offer military support to the Sudanese government in order to secure Chinese shares in Sudan's oil exploitation (Taylor 2007). It is not difficult to understand why China keeps a close eye on Sudan domestic matters, given the fact that CNPC is the largest shareholder of the Greater Nile Petroleum Operating Company, Sudan's largest oil venture. Amnesty International (2006) stated 'China has transferred military, security and police equipment to armed forces and law enforcement agencies in countries where these arms are used for persistent and systematic violations of human rights'. Eventually, wars in Sudan inflicted huge losses on China and forced the country to re-think its 'non-interference' approach. In 2014, the Chinese Foreign Minister Wang Yi made it clear that 'China wanted both sides to stop fighting and seek a reasonable and rational way out.' He also expressed personal willingness to mediate between the warring sides (Wu 2014).

It is true that while some Chinese policies promote peaceful development, others sound militant. Scholars argue that China's intention is always mistakenly assumed to be locking up resources or harming the international energy supply (Garrison 2009). Although these echo certain militant voices in China, such as nationalistic arguments, there are facts that reflect the contrary and are worth further investigating. This does not imply that the West misunderstands China's intent, but studying China's energy security rationale could help in understanding the bigger picture of China's external behaviour, rather than focusing on threats.

The rationale behind China's energy diplomacy

Owing to the energy crisis induced by other countries, energy security has remained a priority on China's agenda since the mid-nineteenth century. It is hard for Chinese leaders to forget both the oil embargo the Coordinating

Committee for Multilateral Export Controls placed during the Korean War in the 1950s and the loss of Soviet oil after the Sino–Soviet split in the 1960s. There is a longstanding energy anxiety among Chinese leaders: 'energy is a key strategic issue for China's economic development, social stability, and national security. As such, China sees energy shortages as one of the biggest potential threats' (Liu 2006).

Although the timely discovery of several giant oil fields made China energy self-sufficient from the 1960s to the 1980s (Chow 1992), anxiety about energy supply rose again in 1994 when increasing energy demand for economic growth had turned China into a net importer of oil. After 1994, one-third of China's total consumption relied on imports and there was no sign of this slowing down. In 2000, China's oil consumption was 4.8 million barrels a day, and this number increased to 8.5 million in 2009, with a current growth of around 900,000 barrels a day (Mouawad 2010). Following the expansion of China's economy, energy demand has rapidly increased rapidly over the past few decades, where the industrial and transport sectors in China are the largest contributors. In 2013, China became the largest oil importer (BBC 2013). According to the IEA's estimate (2012a), by 2035, China's oil demand will rise to almost 15 million barrels per day (mb/d) and its demand for natural gas will increase up to 500 billion cubic metres. As the world's largest energy consumer worldwide, China is facing great pressure to secure its energy supply for rapid economic development and socio-political stability.

Since the 2000s, global competition for resources among powers has become intense and harsh. Within the next few decades, China is expected to face substantially increasing competition for energy, and the security of energy will in turn become more difficult. The necessity of energy security has prompted the Chinese government to impose various energy policies and strategies, both domestically and internationally. While domestic policies tackle issues like energy efficiency and electricity shortages, a remarkable international strategy in the early twenty-first century is international energy cooperation (State Council of PRC 2007). Consequently, China is attempting to pursue energy-related assets abroad, and energy security has become a key component in China's new diplomacy (Breslin 2005).

In its earlier years, energy diplomacy did not constitute an important part of China's overall diplomacy. However, since China turned into an oil-importing country, increasing energy importation and diversifying supply channels have become essential tasks for China's energy-security strategy. China's 11th Five-Year Plan stated that the security of its energy supply should be ensured by 'expanding international energy cooperation', 'actively engaging with the international energy system' and 'making full use of the international market' (State Council of PRC 2001). Energy diplomacy and adopting the going out strategy have thus played a crucial role in China's overall diplomacy, as well as in its energy security. They encourage Chinese energy enterprises to actively invest in resource-rich developing countries in Africa, the Middle East, Central Asia and Latin America.

In order to enhance bilateral energy cooperation, China has adopted energy diplomacy, which is an important part of its 'going out' (State Council of PRC 2011) and national development strategies (Xu 2007). Energy diplomacy is defined as government-involved foreign activities that aim to secure energy resource supply and promote energy-business cooperation (Chen 2008). Zhu Feng (2005), a professor at Peking University and an independent consultant to the Chinese government, argues that energy diplomacy is a logical extension of Chinese national interest, which builds relations with resource-rich countries, develops a favourable environment for Chinese companies in the resource field and establishes alliances for energy cooperation. For China, the ultimate goal of energy diplomacy is to secure national control of overseas resource supplies – particularly oil – and diversify its sources of import. Pang Zhongying (2014), a professor at Renmin University of China and former analyst at the Chinese Embassy in Indonesia, argues that China's energy diplomacy is being carried out as a part component of globalisation. It is not merely only about the market, but it also involves other complicated considerations of geopolitics and strategies. China's former foreign minister, Li Zhaoxin, claimed 'our diplomatic work should provide vigorous support to those efforts aiming to promote international energy cooperation' (Chen 2004). This resonates with other Chinese energy experts who have asserted direct control of overseas energy resources. Chen Huai, an energy expert at the Development Research Centre of the State Council, criticised China's attempt to exploit overseas energy resources with its technology and capital instead of merely purchasing them (EID 2004). Unlike energy trade, which is negotiated among profit-driven firms by cost–benefit analysis, energy diplomacy is not only achieved by intergovernmental agreements, but also through political leverage. Therefore, other than commercial interest, energy diplomacy serves as an important tool to achieve national objectives such as ensuring energy security, managing political risk, expanding international influence and improving inter-state relations (Khan 2008). In other words, while the promotion of energy resource trade is one of the objectives of energy diplomacy, it can also be used to serve other national interests. For instance Ian Taylor (2006), a specialist in African international relations, points out that ideological concerns regarding non-political intervention and the ambition of positioning itself as a global player are parts of China's oil diplomacy in Africa.

Sino–African energy cooperation

The African continent is one of the world's key energy players. It is abundant in energy resources including coal, oil, gas and renewable energy. BP World Energy Statistics indicate that the proved oil reserves in Africa were 10.2 billion tonnes and 13 thousand million barrels in 2002 and 2012 respectively (BP 2013). The difference in this 10-year gap reflects a growing rate in Africa's oil resources. According to the EIA, Africa's proven oil reserves have grown by nearly 120 per cent in the past 30 years. These figures reflect not only the abundance of African natural resources but also their importance to regional economy in Africa. In recent years, while African economic development has had a close relationship with its blooming regional resources economy, the African economy underwent a structural transition through greater diversification to other sectors such as finance, service sectors, transportation, telecommunication and manufacturing. The expansion of multiple sectors other than agriculture and natural resources contributes to the growth in GDP. While the most of foreign capital is attracted by Africa's natural resources, it has also flown into other sectors, like construction, textiles, telecommunications and tourism. According to McKinsey & Company (2010), since 2000, these sectors have generated around 70 per cent of Africa's combined GDP of the four most advanced economies: Egypt, Morocco, South Africa and Tunisia. According to a report from the African Development Bank (2013), the GDP growth rates of African countries are above the global average. Among them, one-third has GDP growth rates of over 6 per cent. Out of the 20 fastest growing countries in the world, 13 are African. Compared to 2011, inflation in African countries has also been slowing down (International Monetary Fund 2012). As a consequence, Africa has become the fastest growing continent in the world. It is expected that investment environment, technology and infrastructure will improve during Africa's economic transition. Although the global financial crisis slowed down the economic growth of African countries, it did not reverse it, as the great development potential in Africa rather drew the attention of foreign investors. New foreign capitals attracted by diversified economic structure would be beneficial to the African economic transition, as well as to regional political stability. Therefore Africa is viewed as 'the second Middle East', where together with a maturing economic environment, growing energy reserves can further create cooperation opportunities in energy sectors.

Following the economic reform in the 1980s, for a decade Africa was neglected in China's foreign policy, which shifted its priority toward Western countries (Taylor 1990). However, this neglect was corrected in the 1990s after the Tiananmen Square protest in 1989, when Beijing realised the importance of working with its Third World allies in order to resist Western sanctions and reinforce energy security. Natural resources imported from foreign sources are substantial to fuel China's economic growth to maintain its political stability. Since Africa is one of very few regions where resources are still available to China, Beijing has rediscovered the continent. Driven by the demand for African resources, China views Africa as part of its going out strategy and has joined the global race for resources in the continent.

China has in recent years increased its global influence by strengthening its national capability through rapid development in political, economical and technological aspects, paving the way for Sino–African energy cooperation. Considering the political instability in the Middle East, strengthening ties with Africa would be an opportunity for China to lower the pressure on its energy security. This is China's global strategy towards the African continent, and this strategy is carried out by establishing bilateral relationships with African countries. Meanwhile, these countries also need to facilitate their energy development through international cooperation, which could bring them financial assistance and technological support. These foreign partnerships could bring considerable revenue, which is important to boost the growth rate in Africa. Through a complementary relationship, China and Africa have in the 2000s made good progress in energy cooperation.

Sino–African cooperation in coal

China and Africa have a long history in coal cooperation and, in the 2000s, this partnership became more mature and diversified. In the field of coal exploitation, China Africa Sunlight Energy Private Limited (CASECO) launched an integrated coal-mining and electricity-generation project in northwest Zimbabwe in 2012 (Xinhua News 2012). CASECO will invest in the mining of coal, the extraction of coal bed methane and the construction of a 600-megawatt power station, as well as being expected to help several upstream and downstream industries, creating over 4,000 job vacancies upon full implementation. In the field of coal investment, Haohua Energy International Hong Kong Resource Cooperation (HEI) reached a deal with Coal of Africa Limited (CoAL) (Mining Weekly 2013), where CoAL agreed to sell the company's shares of US$80 million at 25 pence a share to HEI. With the funds, CoAL will upgrade its processing plant and boost the coal production for international and domestic markets and HEI will become the largest shareholder in the African coal industry. CoAL chairperson David Brown commented: 'this investment by HEI will form the basis for a strategic partnership between CoAL and Beijing Haohua Energy (BHE) that will facilitate the development of CoAL's assets. This is a significant step in stabilising the financial structure of the company and enabling management to unlock the value in the coking assets in the Limpopo province of South Africa.' In the area of coal transportation, China Communications Construction Co. Ltd. signed a two-year contract with the Nacala coal-terminal project in Mozambique, accounting for approximately US$73 million (SASAC 2012). The Nacala coal terminal project is part of Mozambique's Nacala Corridor Project in the northern Mozambican port of Nampula province. Major construction includes principal buildings, a bridge and a rear channel. Implementing the project could promote Mozambique's coal export, regional economy and social development.

Sino–African cooperation in oil and gas

Oil and gas cooperation is traditionally an important pillar in Sino–African cooperation. While Africa has abundant oil and gas resources and large room for development, China has an increasing need of these resources. There are a growing number of Chinese investments in African energy sectors. Direct investments and acquisitions are taking place as Chinese companies start to indirectly invest in resource-rich Western energy companies in Africa through holdings. Regarding recent direct investments, China National Offshore Oil Corporation (CNOOC) has completed the purchase of 66.67 per cent of the Ugandan licences of Tullow Oil plc (2012) with US$2.9 billion. CNOOC has obtained 33.33 per cent of the licence in Exploration Areas 1, 2 and 3A in the Lake Albert Basin and will share the operating responsibilities with other partners. The two parties have been working closely since March 2011 and, based on their timetable, major production should commence in 2016. In mid-2013, China National Petroleum Corporation (CNPC) finalised its purchase of a 28.57 per cent stake in Eni East Africa (Oil and Gas Journal 2013), which before the deal held a 70 per cent interest in the Area 4 gas block in offshore Mozambique. Since then there have been significant gas discoveries in the gas-rich Rovuma Basin offshore Mozambique, so the East African country could in the future become one of the world's largest LNG exporters. CNPC secured its interest with US$4.21 billion to indirectly participate in the Mozambique Rovuma Basin, where Eni has discovered at least 2,124 billion cubic metres of gas. CNPC's entrance into the Mozambique LNG sector is strategically significant for an extension of Sino–African energy cooperation to a new domain. Following the new discovery and production of oil and gas in Africa, China is indeed expected to increase its investment in the continent. Under the existing cooperation framework, the two partners can expand their cooperation dimension to a more stable, effective and mature model. Both sides could utilise a platform such as the Forum on China–Africa Cooperation (FOCAC, see later) in order to enrich their cooperation with more practical functions or new approaches.

Sino–African cooperation in renewable energy

Renewable energy is another highlight of Sino–African cooperation. Since 2012, African renewable energy has undergone rapid development due to governmental support. For example, South Africa has launched a solar water-heater project for 128,000 citizens, has plans to build a 5,000-megawatt power plant as well as to invest US$90 billion in renewable energy over the next 20 years (MOFCOM 2012). Although the African development in renewable energy lacks experience, technological support and regional demand, there is still great market potential with considerable revenue. Therefore transnational companies, including Chinese companies, have expressed interest in entering the African renewable energy market. In 2012, these companies discussed their interest with positive attitudes. Several interested Chinese companies of renewable energy, such as Trony Solar Holdings Co. Ltd., even set up their offices in Africa ahead of time in order to facilitate business expansion to countries like Ghana, Kenya, Nigeria and South Sudan. As they are China's close partners, African countries also welcome Chinese companies. In the

China PV Summit 2012, 18 African delegates visited China to promote investment opportunities in renewable energy in their own countries (China Power 2012). On the other hand, the National Development and Reform Commission (NDRC) also encouraged Chinese solar companies to get on board to invest in Africa. Although Sino–African cooperation in renewable energy is still in its early stage, China will increase its support depending on the pace of demand growth in Africa.

China's advantages in energy investments in Africa

China has become the competitor of Western countries who arrived to Africa a few centuries earlier. Those first on the scene faced many obstacles in the continent and China has taken the advantage to create room to cooperate with African countries. This does not necessarily mean that China has a better and more systematic strategy to take the place of Western countries in Africa, nor that it is doing better than Western companies. This section merely points out several differences between the Chinese and Western business models that have created space for China to subtly enter the market.

First of all, Chinese business models are more welcomed in Africa because China prioritises property over liberal concerns. Horta (2009) argues that 'China's model of a strong government and its focus on economic growth is looked upon by many African despots, and even some democratic leaders, as an example to follow.' Although Western politicians believe Africans are responsible for the corruption and poor governance in the continent, African leaders have been blaming colonialism and 'Western oppression' for its manifold problems (Spillius 2009). Western companies are obliged to operate according to high Western benchmarks with regard to environmental damage, business transparency and human rights. In contrast, Chinese overseas investments always follow the mentality of non-interference in domestic affairs and hence there are no political conditions in business contracts between China and Africa. Although Western involvement draws positive value, such as human rights protection and business transparency, China offer more flexibility in business.

Secondly, China gave Africa a more friendly impression whereby unlike Western countries, this new partner has no history of colonisation, enslavement or military intervention in Africa. While China extracts resources from Africa, it also offers generous aid projects and development assistance. Beyond the most notable Tazam Railway project, Beijing delivers a number of grants, interest-free loans and concessional loans on a bilateral level. Although China prioritises countries such as Angola, Nigeria, Sudan and Zambia, which are rich in resource, Edinger (2008) points out that 'China's approach in terms of development assistance is one of mutual respect where even smaller African countries, with little economic or political significance will receive both aid and investment support. … . It should of course also be noted that China is by no means alone in prioritising development relationships with countries of strategic or commercial significance to itself.'

Thirdly, Chinese labour is more economically competitive than Western labour. Compared to Western labourers and equipment, China's are much cheaper, charging unbeatable prices. As Friedman (2009) notes:

> Coming from richer countries, they insist on better living conditions and bigger profit margins. While Chinese may cluster in Chinatowns and spark a racist backlash, they work harder in harsher circumstances. They accept smaller margins. Their numbers are also infinitely greater than the handful of Euro-Americans moving to Africa. Backed by a government with a seemingly bottomless pool of foreign exchange and a serious commitment to succeeding in Africa, China's energies can transform the continent. The subsidies, cheap money and assistance to business in Africa from the Chinese government will have a huge impact because of the efforts of hard working, globally mobile, entrepreneurial Chinese. Therefore, whereas satisfied European and American investors saw few profit-making opportunities in Africa, entrepreneurial Chinese have found plenty of possibilities.

China's economic reform has created a frenzy of ambition among the Chinese to make profit and send money home by working in Africa. Chinese labourers are very diligent, effective and disciplined but ask for less (such as human rights protection) than Western labourers. They can tolerate living together in packed barracks and seldom complain.[1] Senior staff from Chinese national oil companies in Africa also echoed that the government and companies in Africa welcome the Chinese presence due to their low cost.[2] However, the perception of economically competitive Chinese labour is only believed to be positive at a company or government level, as the Chinese presence in Africa also causes problems to local communities. This will be discussed later.

China's concerns in energy investments in Africa

Although there is extensive room for Sino–African cooperation, there are still a number of obstacles between the two parties. Challenges exist at both domestic and international levels.

Business–cultural differences

There are a number of regional, ethnical and cultural differences between China and Africa. In the first Forum on China–Africa Local Government Cooperation in 2012, Bwansa Mabele, the chief representative of the Congo Business Council in China pointed out that the most significant obstacle in Sino–African cooperation lies in culture (Xinhua News 2012). Owing to cultural differences, both parties fail to engage in smooth communication, and progress of the cooperation is therefore affected. In practice, such differences are reflected in language and business culture. Because of language differences and losses in translation, difficulties unavoidably arise in the exchange of ideas between Chinese and Africans, increasing the cost of cooperation.[3] Regarding business–cultural differences, Chinese businessmen who are influenced by their country's mindset emphasise effectiveness. In contrast, African business culture prefers a freer and looser discipline. Such a difference could create friction and misunderstanding during operations.[4]

Uncertainties in the cooperation model

Different economic capabilities, development situations and regulatory systems are the key consideration in the cooperation model. In the current stage, there is still plenty of room for development in the African energy sector, reflecting not only potential cooperation, but also uncertainties in the cooperation model. In an ideal model, both China and Africa would take the pace and dimension of cooperation into consideration. Cooperation that is either too fast or too broad could damage long-term possibilities and benefits or even result in failing to achieve short-term goals.

Resistance from earlier comers

While Western countries have dominant roles in African energy business, China, as a latecomer, could affect the interests of those who came earlier to the continent. Unavoidably, Western countries would impose resistance and criticism, such as regarding Chinese colonial ambition against Chinese participation in Africa (Edinger 2008). The strategy of Chinese energy companies in Africa is to avoid creating unnecessary conflict with countries that arrived earlier to the scene by 'filling the space' in the African energy market where Western powers are unwilling to or find it risky to invest. The space left by Western companies is an investment opportunity for Chinese companies that are willing to take the risk. At the same time it is like a hot potato, as shown by CNPC's experience in Sudan.

Local competition

China is not the only player in the African energy sector. For example, in the aforementioned section, CNPC is a newcomer in the Mozambique LNG industry. After CNPC's investment, Eni, an Italian energy company, remains the indirect owner of 50 per cent of the participation in gas block Area 4. Other than these two major players, Empresa Nacional de Hidrocarbonetos from Mozambique, Kogas from Korea and Galp Energia from Portugal evenly share the remaining 30 per cent of participation. Companies from Japan, Thailand and Malaysia are also operating in other gas blocks in the region. This does not mean that foreign companies would resist China's involvement. In Sino–African cooperation, energy projects are usually joint ones between different countries. In order to facilitate the cooperation, China should not only focus on its partner, Africa, but also on other factors imposed by other participants in the continent.

Local conflict

Security is a concern for all investors in Africa, and the conflict in South Sudan is a particular headache for China (Smith 2014). Since the independence of South Sudan in 2011, conflicts have broken out from time to time and oil has been a potential spark. Although China stubbornly persists in its non-intervention doctrine, the conflicts in South Sudan have forced China to impose a rare overt political intervention in the region, calling for an immediate end to hostilities and protection of vital oilfields from rebels. Clashes between rebel forces and the South Sudanese government have caused over 1,000 fatalities and have reduced oil flows by 20 per cent. Given that China is the single biggest trading partner in the region, the Chinese foreign minister Wang Yi met both delegations from the rebels and the government to help restore stability.

Somali piracy

Another security concern is Somali pirate attacks on vessels and offshore facilities, for example in the case of Mozambique LNG exports that take up to 10 days for long-haul shipping. Therefore, like other oil-producing countries in Africa, Mozambique LNG vessels are facing the threat of pirate attacks. In addition, the scope of activities of Somali pirates has been extending to the waters south of Mozambique, and offshore natural gas facilities in the region are exposed to these attacks. Somali piracy is a problem at an international level; according to the International Maritime Bureau, 264 attacks by Somali pirates were recorded worldwide in 2013. Owing to the strengthening of international efforts to fight Somali pirates, the number of pirate attacks has declined in recent years. However, using armed forces to protect offshore facilities and vessels from pirate attacks would result in extra production costs and expansive insurance.

Local grievance towards Chinese presence

The Chinese presence in Africa is not always idyllic. Thousands of Chinese migrants moving into Africa have become the source of social problems in the continent. Although the Chinese have limited contact with the local population, grievance towards them is unavoidably aggravated. Not only are hard-working Chinese labourers reducing job opportunities for local Africans, Chinese street sellers are also endangering local businesses.

Chinese products that are exported to African countries are also damaging the local industries. Moreover, many of the Chinese migrants in Africa are illegal, therefore harming China's image as a great power. Africa questions why such a great power would send thousands of people to a poor country to compete with even poorer locals who have been there for generations. Growing resentment in the African society has resulted in conflicts and violent incidents.

A way to global energy governance?

Within the next 40 years, the energy demands of the world will continue to rise and gas demand is expected to grow at a significant rate. The bulk of this growth in gas demand will be led by 'emerging Asia', including China. Asian countries need to import more natural gas and LNG therefore has an increasingly important role in the global supply of energy. Asian gas buyers currently rely on expensive Australia LNG and limited Middle East LNG. They are willing to import cheaper LNG from more distant regions such as the Gulf of Mexico and Africa. Therefore as long as the price is right, African LNG is always desirable. Together with the driving force of the US shale-gas revolution, the trend of global energy markets is shifting to the East Asian market. Oil and gas producers in Africa have begun to adjust their energy strategies in order to grab the opportunity created by the increasing demand of Asia. Abdalla S. El-Badri, the OPEC Secretary General, pointed out in a conference that since oil and gas will continue to dominate the global energy market in the foreseeable future, the security of oil and gas prices in the world economy is very important. Even though countries are developing low-carbon economies, global oil and gas consumption will continue to rise because of the demand of oil as transportation fuel in developing countries and of natural gas as a replacement for coal in power generation. Under such circumstances, Abdalla S. El-Badri emphasised stabilisation of the energy market as the key to global energy security, echoing Yergin's (1988) definition of energy security: 'assur[ing] adequate, reliable supplies of energy at reasonable prices and in ways that do not jeopardise major national values and objectives'.

The scale, size and complexity of the energy market make it a unique product of oil and gas, and the energy price stability of each country is closely related. In the international energy market, neither a high price nor a low price is good in the long-term for either producer or consumer; equilibrium is important. Although no-one can ensure that long-term energy prices do not fluctuate, countries can try to develop a market framework to improve the clarity and predictability of the energy market. The energy dialogue between producer and consumer countries needs to be strengthened through international platforms, and global energy governance is needed. In recent years, countries have been discussing how to avoid dramatic price fluctuations, which is crucial to improving the function of the oil and gas market.

Since 2011, the International Energy Forum (IEF), the International Energy Agency, OPEC and other international organisations have started to cooperate on how to improve energy-market transparency and energy-outlook association between the data and the oil market. It has also been suggested that energy issues be brought onto a broader international platform such as the G20.[5]

Other than a top-down approach, regional energy governance could contribute via a bottom-up channel. The Forum on China-Africa Cooperation (FOCAC), first held in 2000, serves as a platform for Chinese and African leaders to coordinate more closely to promote energy cooperation, as stated in FOCAC's fifth action plan: 'in view of the strong complementarity and cooperation potential between China and Africa in energy and resources, the two sides will encourage and support joint development and proper use of their energy and resources by enterprises of the two sides. They will consider the establishment of a China-Africa energy forum under the framework of FOCAC to promote China-Africa energy exchanges and cooperation.' China's energy cooperation with Africa could strengthen regional energy security, which could in turn stabilise the regional energy market. According to the U.S. Department of Energy, although China's vast investment in energy assets abroad draws concerns, its effect is economically neutral. In fact, China's going-out strategy could let Chinese energy companies explore energy reserves in African regions where no Western powers could or would invest, and over 50 per cent of China's overseas oil production in 2008 was sold outside of China (Dittrick 2010). This increases the world's available energy reserves and stabilises the energy market. Therefore instead of harming global energy security, China's energy policy could actually enhance it.

In short, if China wishes to secure its energy supply, then the key is in stabilising the global market. One way to do so is by promoting international energy cooperation and, as discussed above, Sino–African energy cooperation is one of the relationships China needs to enhance. Obstacles include a number of domestic and international concerns, but of the major challenges one is how to fit such a relationship into the international energy system. Through a multilateral approach, global energy governance could offer the leverage to boost Sino–African energy cooperation. The momentum here could contribute to global energy security, from which China may eventually benefit.

Notes

1 Personal conversation with senior researchers from national oil companies.
2 Personal conversation with overseas staffs from CNOC based in Africa.
3 Personal conversation with senior researchers from national oil companies.
4 Personal conversation with overseas staffs from CNOC based in Africa.
5 Personal conversation with senior officers from energy-related government departments.

Bibliography

African Development Bank Group (2013) *Annual Development Effectiveness Review 2013: Towards Sustainable Growth for Africa.* Available at: www.afdb.org/fileadmin/ uploads/afdb/Documents/Project-and-Operations/ADER-%20Annual%20Developm ent%20Effectiveness%20Review%202013.pdf (accessed 29 January 2014).

Amnesty International (2006) *People's Republic of China: Sustaining Conflict and Human Rights Abuses: the Flow of Arms Accelerates.* New York: Amnesty International.

Baghat, G. (2006) 'Europe's Energy Security: Challenges and Opportunities.' *International Affairs* 82(5): 961–975.

Bajpaee, C. (2005) 'China Fuels Energy Cold War', *Asia Times*, 2 March, 1–8. Available at: www.atimes.com/atimes/China/GC02Ad07.html (accessed 28 January 2014).

BBC (2013) 'China Overtakes US as the Biggest Importer of Oil', *BBC*, 10 October. Available at: www.bbc.co.uk/news/business-24475934 (accessed 29 January 2014).

BP World Energy Statistics (2013) Available at: www.bp.com/content/dam/bp -country/fr_fr/Documents/Rapportsetpublications/statistical_review_of_world_energ y_2013.pdf.

Breslin, S. (2005) 'Power and Production: Rethinking China's Global Economic Role.' *Review of International Studies* 31: 735–753

Calder, K. (2006) *Pacific Defense: Arms, Energy, and America's Future in Asia.* New York: William Morrow and Co.

Chen, J. (2004) 'The Five-Circle Energy Diplomacy of China', *21st Century Business Herald*, 18 March. Available at: www.21cbh.com/HTML/2004-3-18/14529.html (accessed 30 January 2014).

Chen, M. and Jaffe, A. M. (2007) 'Energy Security: Meeting the Growing Challenge of National Oil Companies.' *The Whitehead Journal of Diplomacy and International Relations*, Summer/Fall: 9–21.

Chen, S. (2008) 'Motivations Behind China's Foreign Oil Quest: A Perspective From the Chinese Government and the Oil Companies.' *Journal of Chinese Political Science* 13(1): 79–104.

China Power (2012) '18 African Countries Visit China to Promote Solar Energy', *Chinapower*, 19 December. Available at: www.chinapower.com.cn/article/1219/a rt1219167.asp (accessed 29 January 2014).

Chow, L. (1992) 'The Changing Role of Oil in Chinese Exports, 1974–1989.' *China Quarterly* 131: 750–765.

Dittrick, P. (2010) 'Chinese Oil Companies Invest Heavily Abroad.' *Oil and Gas Journal* 108(5).

Edinger, H. (2008) *Colonial Ambitions?* Available at https://newmatilda.com/2008/08/ 11/colonial-ambitions (accessed 29 January 2014).

EID (2004) 'Chinese Oil Firms "Go Out" to Resolve the Oil Predicament', *Economic Information Daily*, 21 September. Available at: http://finance.sina.com.cn/g/ 20040921/08431037218.shtml (accessed 29 January 2014).

Friedman, E. (2009) 'China-Driven Development as China Pours Billions Into Africa, Other Countries Are Trying to Keep Up', *Beijing Review*, 5 February. Available at: www.bjreview.com/world/txt/2009-02/01/content_176304.htm (accessed 29 January 2014).

Garrison, J. (2009) *China and the Energy Equation in Asia: The Determinants of Policy Choice*. First Forum Press.

Horta, L. (2009) *China and Africa*. Available at: www.asiasentinel.com/index.php?op tion=com_content&task=view&id=2154&Itemid=422 (accessed 29 January 2014).

IEA (2012) *Oil & Gas Security Emergency Response of IEA Countries People's Republic of China*. Paris: IEA, pp. 1–19.

International Monetary Fund (2012) *Regional Economic Outlook: Sub Saharan Africa Maintaining Growth in an Uncertain World*. Available at: www.imf.org/external/p ubs/ft/reo/2012/afr/eng/sreo1012.pdf (accessed 29 January 2014).

Karon, T. (2006) 'Iran Diplomacy: Why Russia and China Won't Play Ball', *Time*, 22 May, 1–2. Available at: www.time.com/time/world/article/0,8599,1175573,00.html (accessed 28 January 2014).

Khan, H. (2008) 'China's Energy Drive and Diplomacy.' *International Review* (Shanghai Institute of International Studies): 93–94.

Langenkamp, D. (2010) *Our Friend, The Dragon*. Available at: http://207.41.118.84/a rticles.cfm?aid=3758 (accessed 29 January 2013).

Lim, R. (2005) *Geopolitics of East Asia: The Search for Equilibrium*. New York: Routledge.

Liu, X. (2006) 'China's Energy Security and Its Grand Strategy.' *The Stanley Foundation Policy Analysis Briefs*: 1–16.

McKinsey & Company (2010) *What's Driving Africa's Growth?* Available at: www. mckinsey.com/insights/economic_studies/whats_driving_africas_growth (accessed 29 January 2014).

Mearsheimer, J. (2006) 'China's Unpeaceful Rise.' *Current History* 105(690): 160–162.

Mearsheimer, J. (2010) 'Australians Should Fear the Rise of China.' *The Spectator* (October). Available at: http://mearsheimer.uchicago.edu/pdfs/A0053.pdf (accessed 29 January 2014).

Mining Weekly (2013) 'CoAL Shareholders Approve $100m HEI Investment', *Mining Weekly*, 25 January. Available at: www.miningweekly.com/article/coal-shareholder s-approve-100-million-investment-by-hei-2013-01-25 (accessed 29 January 2014).

MOFCOM (2012) *Enormous Potential in Africa's Solar and Other New Energy Markets*. Available at: http://eg.mofcom.gov.cn/article/jmxw/201212/2012120849 0225.shtml (accessed 30 January 2014).

Mouawad, J. (2010) 'China's Rapid Growth Shifts the Geopolitics of Oil', *New York Times*, 20 March.

Mufson, S. (2006) 'As China, U.S. Vie for More Oil, Diplomatic Friction May Follow', *Washington Post*, 15 April, 1–4. Available at: www.washingtonpost.com/ wp-dyn/content/article/2006/04/14/AR2006041401682_2.html (accessed 28 January 2014).

Oil & Gas Journal (2013) *CNPC Completes Buy of Stake Off Mozambique*. Available at: www.ogj.com/articles/2013/07/cnpc-completes-buy-of-stake-off-mozambique.html (accessed 30 January 2014).

Pang, Z. (2012) 'Peaceful Development and Energy Diplomacy.' *Outlook Weekly* 95.

SASAC (2012) *China Communications Construction Co Ltd Signed Nacala Coal Terminal Project Master Contract in Mozambique*. Available at: www.sasac.gov.cn/ n1180/n1226/n2410/n314289/14825595.html (accessed 30 January 2014).

Ross, M. (2008) 'Blood Barrels: Why Oil Wealth Fuels Conflict', *Foreign Affairs*, May/June. Available at: www.foreignaffairs.com/articles/63396/michael-l-ross/ blood-barrels.

Smith, D. (2014) 'China Urges Immediate End to Conflict in South Sudan', *The Guardian*, 6 January. Available at: www.theguardian.com/world/2014/jan/06/presi dents-sudan-south-sudan-meet-juba-discuss-conflict (accessed 28 January 2014).

Spillius, A. (2009) 'Barack Obama Tells Africa to Stop Blaming Colonialism for Problems', *The Telegraph*, 9 July. Available at: www.telegraph.co.uk/news/world news/africaandindianocean/5778804/Barack-Obama-tells-Africa-to-stop-blaming-co lonialism-for-problems.html (accessed 1 September 2014).

State Council of PRC (2001) *China's 11th Five Year Plan*. Available at: http://politics. people.com.cn/GB/1026/4208451.html (accessed 23 December 2012).

State Council of PRC (2007) *Zhongguo nengyuan zhengce* [*China's Energy Situation and Policy*]. Available at: www.gov.cn/zwgk/2007-12/26/content_844159.htm (accessed 29 January 2013).

State Council of PRC (2011) 'An Overview of "Going Out" Strategy.' Available at: http://qwgzyj.gqb.gov.cn/yjytt/159/1743.shtml (accessed 23 December 2012).

Taylor, I. (1990) 'China's Foreign Policy Towards Africa in the 1990s.' *Journal of Modern African Studies* 26(3): 443–460.

Taylor, I. (2006) *China and Africa: Engagement and Compromise*. London: Routledge.

Taylor, I. (2006) 'China's Oil Diplomacy in Africa.' *International Affairs* 82(5): 937–959.

Taylor, I. (2007) 'Unpacking China's Resource Diplomacy in Africa.' In M. C. Lee, H. Melber, S. Naidu and I. Taylor (eds), *China in Africa*. Vällingby: Elanders Gotab AB.

The White House (2006) 'The National Strategy.' Available at: www.comw.org/qdr/ fulltext/nss2006.pdf.

Tullow Oil plc (2012) '$2.9bn Farm-Down of Uganda Licences Completed.' Available at: www.tullowoil.com/index.asp?pageid=137&newsid=737 (accessed 30 January 2014).

Walt, S. (2011) 'Sino-American Rivalry: A Chinese View.' *Foreign Policy* (21 November). Available at: http://walt.foreignpolicy.com/posts/2011/11/21/sino_am erican_rivalry_a_chinese_view (accessed 29 January 2013).

Wu, Y. (2014) 'China's Oil Fears Over South Sudan Fighting', *BBC*, 8 January. Available at: www.bbc.com/news/world-africa-25654155 (accessed 29 January 2013).

Xinhua News (2012) 'China and Zimbabwe Cooperate to Develop Coal Resources.' Available at: http://news.xinhuanet.com/world/2012-12/20/c_124123320.htm (accessed 29 January 2014).

Xinhua News (2012) 'Rapid Development in Sino-African Bilateral Trade.' Available at: http://big5.xinhuanet.com/gate/big5/www.bj.xinhuanet.com/zt/2012-09/02/c_112 930335.htm (accessed 29 January 2014).

Xu, Q. H. (2007) 'China's Energy Diplomacy and Its Implications for Global Energy Security.' *FES* Briefing Paper. 1–8.

Yergin, D. (1988) 'Energy Security in the 1990s.' *Foreign Affairs* 67(1): 110–132.

Yergin, D. (2006) 'Ensuring Energy Security.' *Foreign Affairs* 85(2): 69–82.

Zhu, F. (2005) 'A High Price to Pay: China's Resource Diplomacy Requires Wisdom.' *New Finance* (18 May). Available at: http://media.163.com/05/0518/10/1K1FC 60A00141A16.html (accessed 29 January 2014).

Zweig, D. (2009) 'A New "Trading State" Meets the Developing World.' Working Paper No. 31. *Center on China's Transnational Relations*.

Zweig, D. (2010) *China's Energy Rise, the US, and the New Geopolitics of Energy*. Los Angeles, CA: Pacific Council on International Policy.

Zweig, D. and Jianbai, B. (2005) 'China's Global Hunt for Energy.' *Foreign Affairs* 84(5): 25–38.

9 Decentralised energy resources as a challenge to the secular top-down paradigm of energy production?

Marie-Hélène Schwoob

In the past, energy was a matter of the central government: importing countries were supporting the energy supply from abroad through the establishment and maintenance of good diplomatic relations, whereas domestic energy supply was taken care of by state-owned enterprises. China was not exempt from this model and its energy supply policies are still mainly featured in this way. However, in addition to an ever-growing demand, its energy strategy must today face new rising concerns, such as difficulties in developing and adapting energy and electricity networks and the increasing threat of environmental issues – above all, air pollution. In an attempt to answer these questions, local stakeholders build small and clean energy production sites fed by renewable resources near consumption areas. However, these recent developments of new energy sources are a challenge to the country's traditional top-down political model. Decentralisation and privatisation reforms, though regularly pushed by the government, are hampered by stakeholders' tactics to keep on maintaining control over energy production.

China's strategy of preferring centralised energy production to secure energy supply

How the 'central model' developed in China: the birth of the NOCs

The reforms China has been experiencing since 1978 have led to an exceptional economic development, as the country has managed to maintain almost 10 per cent of the annual average economic growth for at least 20 years. However, the 'Chinese miracle' had its downsides, a tremendous rise in domestic energy demand among others. Answering this rise was – and still is – an important obligation to feed the country's industrial development, which remains at the heart of Chinese economic growth. Moreover, responding to the increasing energy demand is also strongly linked to social and political stability,[1] the core issue for the Chinese leadership.

The central government, challenged by the need to provide efficient answers to the country's rising demand, took charge of the arduous task of improving the national supply, which eventually led to the creation of the three big

National Oil Companies in the early 1980s. In the initial years of the economyic reform, which started after Deng Xiaoping's accession to power, the idea was to radically transform the administrative system and evolve towards a more market-driven economy. Chinese institutions inherited from the Maoist period were indeed characterised by significant overlaps of economic and political functions: in 1970, there were no less than 100 ministries and commissions in charge of decision-making at the central level (Waldron *et al.* 2006: 282). Administrative entities were handling functions linked to economic production and had ownership rights over machinery, factories, mines and other productive assets, then from 1978 onward progressively transferred to (the newly created) state-owned enterprises. During this period, the government was conferred control over economic activities through the possibility to establish less direct management mechanisms (e.g. through regulation).

In 1982 in the field of energy, a certain amount of tasks that had previously been performed by the Ministry of Petroleum Industry were transferred to a new body, the China National Offshore Oil Corporation (CNOOC), which was specifically created to engage in offshore joint ventures projects (Lieberthal and Oksenberg 1988: 86). In 1983, productive assets of the Ministry of Petroleum Industry, the Ministry of Chemicals and the Ministry of Textile were transferred to the China Petrochemical Corporation, better known as Sinopec. Finally, in 1988, the remaining production activities once held by the Ministry of Petroleum Industry were reassigned to another entity, the China National Petroleum Corporation (CNPC). The initial objective of the creation of these energy SOEs and of transferring the energy production tasks from ministries to market-ruled bodies was to promote competition and market rules, in line with the new economic ideas adopted through the reforms. Increased competition among energy suppliers was thus supposed to lead to better economic efficiency. In addition, the state enabled non-state actors to enter upstream sectors (e.g. in coal mining) in order to increase the state energy output, which in the early 1980s was stagnating (Naughton 1995: 224).

However, the energy value chain was in fact carefully divided among Chinese NOCs, showing that the way they 'got' their share in the national energy market still heavily inherited centrally planned patterns of political economy. Whereas the CNOOC was set up to look for international upstream assets to feed domestic production sites and was given downstream activities, such as refinery, the CNPC was put in charge of building and operating pipelines. The three NOCs were also initially geographically divided. While CNPC was controlling northern areas of China, Sinopec was dominant in the South. As for the CNOOC, it was assigned the task of supervising offshore assets. In other words, though the entities dealing with energy production, transportation and diffusion changed 'face', energy administration essentially kept the central planning features despite the willingness to proceed to market reforms.

Market reforms 'lacking market characteristics'

The successive rounds of market reforms, which took place in the 1980s and 1990s, ended with the rigid division of tasks and areas previously attributed to NOCs and gave them the possibility to move the value chain upstream or downstream and to compete for assets beyond the geographical area they had originally been allocated to. However, despite the possibility of expanding beyond the principal granted activities, the division of roles among NOCs is still a constant in the Chinese energy market, keeping the CNPC as the main player in pipeline construction and operation, Sinopec as the largest oil refiner and the CNOOC as the leader of offshore upstream production (Dan 2007: 116). In spite of market reforms, the Chinese NOCs thus stayed in quasi-monopolistic situations in their original fields of activity.

At the beginning of the 2000s, all of the three NOCs undertook market capitalisation, which led to listing their subsidiaries in Hong Kong's stock exchange and enabled enterprises to raise several billion US dollars. However, this move towards market capitalisation did not greatly alter their situation. First, parent companies own the majority of shares of their subsidiaries (the CNPC holds more than 86 per cent of PetroChina, Sinopec holds more than 75 per cent of Sinopec Corp., and the CNOOC holds more than 66 per cent of CNOOC Ltd. (Downs 2010: 75)). In addition, despite the diversification of shareholders, the Chinese state managed to maintain levers of control over their activities (investments have to be approved by the Party), personnel (the Party can appoint, dismiss or promote the companies' general managers) and financial resources (through low interest loans provided by State banks). Finally, the NOCs still take benefit from their hierarchical status at the ministerial level, inherited from the former ministries they grew out of. The strong links maintained between the government and the NOCs enabled them to preserve their dominant position in the energy market.

Faced with the necessity to speed up market reforms, the government regularly tried to encourage private actors to incorporate the energy sector as well. However, the prerequisites required to enter the energy market were too difficult to fulfil for private enterprises (in terms of level of investments, access to distribution networks, etc.), especially regarding upstream activities, which are still dominated by the big NOCs. As Shi Dan (2007: 116) sums it up, 'in the oil processing and coking industry, over 80 per cent of actors are non-state but the shares of their capital, output and added-value are of a mere 15.33 per cent, 14.60 per cent and 19.13 per cent respectively'.

A state-enterprises nexus

Despite the announced market reforms, the energy administration model that emerged from the creation of NOCs was characterised more by a centralised economy approach and a strict control over energy resources in the hands of the central government. The domination of the energy sector by the three big national companies was in fact an opportunity rather than a challenge for the Chinese state. Built on former ministries, the NOCs maintained strong ties with central institutions, which was not always possible with private entrepreneurs who can easily by-pass governmental control. The State-owned Assets Supervision and Administration Commission (SASAC) indeed has formal authority over NOCs, in the same way it has over other Chinese SOEs. Moreover, national energy companies also keep very close relationships with the National Energy Administration (NEA) – the supreme energy administration organ within the central planning institution, the National Development and Reform Commission (NDRC) – as well as with other ministerial-level bodies. Consequently, the fact that NOCs were created on the remaining parts of former ministries not only provided them with inherited links with the government,[2] but also awarded them the same administrative status as their counterparts within central bodies, thus a right to exchange on an equal basis.

Links between the NOCs and central institutions are not only a consequence of shared interests between stakeholders. Human resources also circulate between the two entities, as Erica Downs (2007: 70–71) points out:

> Some of China's senior leaders previously worked in the oil industry, including Zeng Qinghong (CNOOC), a member of the Politburo Standing Committee and China's Vice President, Zhou Yongkang (CNPC), a member of the Politburo and a Vice Premier of the State Council. Some employees of China's national oil companies similarly spent time earlier in their careers working for the Chinese government, such as former CNPC Vice President Wu Yaowen, who had worked for both the Ministry of Energy and the State Planning Commission, and the new president of CNPC and chairman of PetroChina, Jiang Jiemian, who previously was deputy governor of Qinghai Province.

There are several rationales behind the domination of the energy sector by a small number of SOEs strongly tied to central bodies (which was also the preferred strategy of European countries just a few years ago[3]). First, energy security is a national stake and in the case of China, as well as of other states, answering to energy demand is key for economic growth and social and political stability. It is thus perfectly understandable that states have generally been reluctant to give full responsibility of this mission to fragmented and non-coordinated private stakeholders, subject to market fluctuations and business difficulties. A second important reason is that energy security is strongly correlated to international relations. The energy crisis of the 1970s and 1980s not only demonstrated the necessity and importance of securing energy supply, but also proved that controlling energy production sites and energy channels could provide countries with powerful tools for coercion. Energy supply does not only rely on the market rules of a globalised world, like other traded products, but its strategic importance makes it a product that is tradable on the basis of relationships between countries, thus relying on issues beyond prices and availability. This characteristic prevents the possibility of seeing an energy market entirely based on market rules. The diplomatic intervention of states remains necessary – especially in a context where more and more countries are competing for resources[4] and are requiring a renewed 'international energy governance'. Last but not least, restricting the energy sector to a limited number of energy suppliers enables a concentration of resources, which can be highly useful in the early development stages. The concentration of financial resources indeed enables stakeholders to channel large investments towards building heavy electricity transmission and distribution infrastructures, as well as towards research and development (e.g. for nuclear energy), both keys to answering increasing demand.

Thanks to this configuration of energy supply management established through the reforms at the beginning of the years 2000, energy resources from abroad were supported by diplomatic relations and operated by the CNOOC, whereas domestic energy supply was taken care of by the CNPC and Sinopec, responsible for creating centralised production sites that feed the whole country. However, new challenges for the Chinese energy administration appeared on the eve of the twenty-first century.

The emergence of decentralised energy sources as new solutions to environmental issues and problematic energy supply in remote areas

New stakes at hand calling for a development of clean energy

The first challenge experienced by authorities at the beginning of the 2000s was an acceleration of the rise in demand, which started leading to serious repeated energy shortages. In 2003, shortage of energy seriously disrupted industrial activities and forced the country to raise its imports. In 2004, 2005 and 2008, the country again had to face an electricity shortage that forced the country to try to diversify its energy sources, not only geographically but by per type of resource as well. Driven by strong economic growth, in those years the country's energy demand was indeed increasing four times faster than it had during the past decades.[5] This unprecedented growth brought China to become, in 2010, the first world energy consumer, ahead of the USA. Years later, the trend that emerged in those years is not likely to slow down. According to the estimations of the International Energy Agency (IEA), China's energy demand will rise by 75 per cent between 2008 and 2035, at which point China will be able to consume 70 per cent more energy resources than the USA (IEA 2011), even if energy consumption per capita will remain half that of a US citizen's.

At the beginning of the twenty-first century, challenges regarding energy consumption were not only linked to the need to secure supplies. Beyond traditional energy security concerns, environmental problems became another important challenge brought on by energy choices, a question the country could no longer choose to ignore. Above all, air pollution caused by accelerated industrial development and rapid urbanisation, together with the peculiar coal-dominated structure of China's energy mix, started to have its choking effects on cities (see Romano, this volume, Chapter 11). In 2008, the country became the first greenhouse gas emitter, and according to IEA's forecasts, the rise of coal consumption will continue at least up to 2020. Carbon dioxide emissions are thus expected to double by 2030 and account for one third of worldwide emissions, even if per capita emissions will remain below the global average. However, beyond the effects that the use of coal has on global climate, air pollution also has an important and direct cost on public health: according to the WHO's database, China would have the highest rate of deaths attributed to outdoor pollution, far ahead of the second-ranked country, India (470,649 against 168,601 in 2008). Environmental issues in general would be responsible for 3–12 per cent of GDP annual losses.[6]

Although air pollution was not new for the country, the levels of pollution recently reached started posing a serious threat to social stability, especially in urban areas. In January 2013, a thick haze hovered over eastern China and enveloped the capital. In October 2013, an impenetrable smog (1,000 micrograms of PM 2.5 per cubic meter[7]) covered the city of Harbin and completely paralysed the transportation systems (air and road traffic), as well as the economy. China's young urban middle class took over social networks so as to vent their anger and disappointment. People also started engaging in concrete actions. In 2011, for example, a thermal plant project was brought to a halt because of protests in Haimen (Jiangsu).

Aware of the dramatic proportions these issues could take on, the central government, in its 10th Energy Five-Year Plan (2000–2005), introduced two new pillars to the country's energy strategy[8]: the first one aims to improve energy efficiency and the second one aims to alleviate problems posed by the coal-dominated energy mix (targets were set for the share of clean energy in the energy mix). In the 11th Five-Year Plan (2006–2011), more ambitious targets were promulgated: energy intensity had to improve by 20 per cent by 2010. In 2009, in the framework of UNFCCC negotiations, the State Council also announced that China was going to reduce CO_2 per unit of GDP by 40–45 per cent by 2020, relative to 2005. Targets for the 12th Five-Year Plan pledge that China will increase the proportion of non-fossil fuels in overall primary energy use to 11.4 per cent by 2015, and to 15 per cent by 2020. In order to reach its objectives in terms of GHG emissions and to diversify its energy mix, the country has decided to develop alternative sources of nationally produced clean energy. In line with this strategy, wind and solar power, among others, have experienced the development of an impressive range of incentive policies.

Wind and solar decentralised capacities answers to both environmental issues and energy security in remote areas

The Renewable Energy Law (REL), promulgated in 2005 (and amended in 2009), played a major role in the development of these resources. The REL is actually more of a comprehensive 'programme' than a law: it assigns targets for installed production capacities, sets up incentive policies (special funds, subsidies, tax abatements) as well as economic mechanisms, which apply to all of the actors in the sector (feed-in tariffs, cost sharing, mandatory buying). The document shows that China has great ambitions in terms of renewable energy development. Indeed, by 2015, solar power capacities should reach 15 gigawatt, whereas for wind power, targets were set up at 50 gigawatt. By 2020, development objectives are respectively 90 gigawatt and 150 gigawatt. In total (i.e. including other renewable energy sources), renewables should account for 11.4 per cent of China's energy mix by 2015, and for 20 per cent by 2020.[9]

The whole set of policies introduced by the REL led to a rapid development of renewable energy production capacities. Wind power was the first to feel the effects of their introduction: from nearly zero in 2005, installed capacities expanded beyond 10 gigawatt in 2008. Then, from 2008 to 2010, wind power capacities more than doubled each year, and China's overall capacity went beyond 60 gigawatt in 2011 (China Renewable Energy Industry Association 2012). Solar energy development demonstrated less enthusiasm: installed capacities accounted for less than 1 gigawatt in 2010, particularly because of the unwavering high costs of technologies.

Besides the fact that wind and solar power are nationally produced and do not emit greenhouse gases, they are also at an advantage due to their decentralised nature. Wind and solar power indeed enable the development of small and decentralised plants that produce energy close to where it will be used – and partially solving the issue of energy shortages. This 'close-to-customers' configuration is in line with the precepts of the concept of 'energy transition', responding both to concerns in energy supply and to environmental and efficiency issues. Customers who enjoy the services of decentralised energy production bases would not have to share a supply or rely on a limited number of large and remote power stations, but would rather benefit from energy sources produced locally that would also significantly reduce energy waste. Such a configuration could not only partly alleviate the serious shortages that the Chinese electricity system has been facing over the past few years (both in urban eastern areas and in western remote regions), but could also contribute to reducing the polluting emissions coming from coal combustion. China's coal resources are certainly abundant, which makes this resource still account for almost 80 per cent of electricity production. However, coal production areas are located far from those of consumption. The transportation system (mainly train travel, which 70 per cent of coal transport relies on) could not meet the growth of

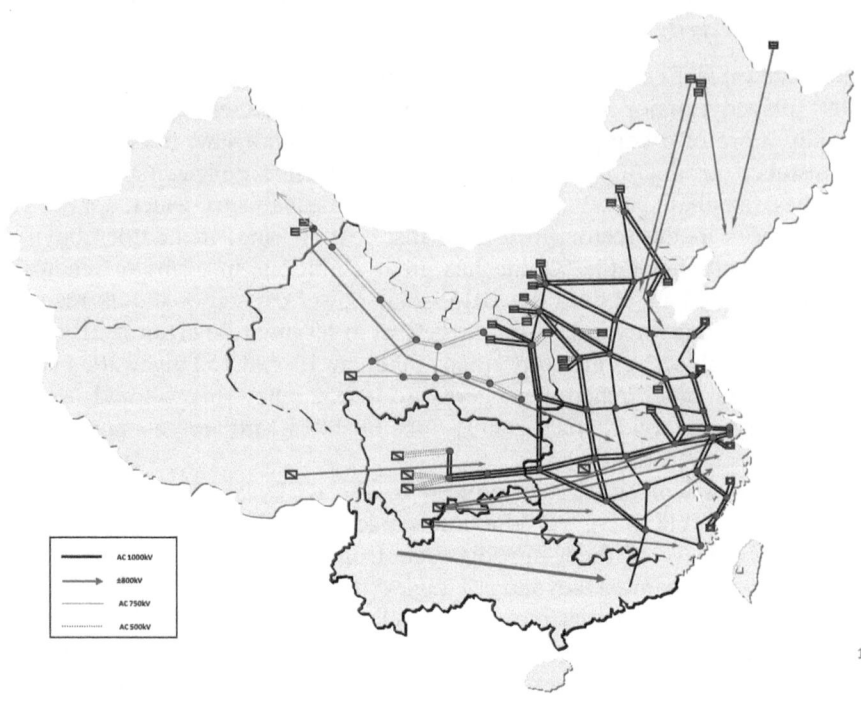

Figure 9.1 National grid by 2020
Source: Adapted from China Electric Power Research Institute.

production capacities, and is regularly clogged by the considerable quantities of coal that transit from the west to the east (Schwoob 2011).

Against this background, the development of decentralised, renewable energy plants could help in alleviating the supply issues experienced by remote areas that are located far from coal mines and thermal plants and that are still not connected to the grid. Linking these to energy networks is complicated due to the current situation of coal transportation throughout China. An alternative would thus be to connect them to electricity grids. However, since installing new electricity lines is very expensive (more than US$100,000 per kilometre), linking remote areas to traditional energy grids would significantly lower the profitability of projects, especially if the number of consumers at the end of the line is not significant. From this point of view, it is clear that the use of renewable energy would constitute a revolutionary way to produce and distribute electricity, in the sense that it might rethink the traditional and highly centralised model of energy production and bring solutions to reducing transmission losses, to lowering carbon emission and to generating increasing economic benefits.

Reforms and roadblocks: decentralised energy sources facing the traditional centralisation model of energy policies

Incentive mechanisms as tools for the central State to keep control over their development

Wind and solar power capacities can take various forms. At one extremity of the spectrum there are very large-scale wind and solar farms: in 2012, China's GCL-Poly Energy Holdings Ltd., the world's largest producer of poly-silicon,[10] gained approval by the National Energy Administration to build the world's largest solar farm in Datong. As for wind power, the 11th Five-Year Plan initially wished to develop seven mega-plants, which are supposed to reach 148 gigawatt of installed wind power capacity by 2020. At the other extremity, there is a multitude of small plants producing energy from solar or wind resources.

The first model is consistent with the traditional way of producing energy: large plants provide electricity, which is then sent to consumers through the national grid. This configuration would thus confirm the classical centrally led and centrally owned model of energy production and transmission. The development of the second model, on the other hand, faces certain obstacles. It is clear that the proliferation of small and decentralised energy plants found outside of central state's areas of control do not gain governmental support. The Renewable Energy Law explicitly sets up mechanisms that *de facto* allow the central state to exercise control over the development of production capacities, in some way 'taming' the opportunities for these sources of energy to develop according to local needs.

One of the main ways the government 'regulates' their development is through feed-in tariffs. In a context where renewables are far less profitable than traditional energy sources due to their low level of technological readiness, feed-in tariffs for electricity prices are essential to encourage investors to engage in the sector. According to regulations, in order to benefit from them, producers have to gain national permits and authorisations. Grid companies are then supposed to buy renewable energy at a certain price, set up by the NDRC. In 2009, feed-in tariffs were chosen for wind power: the tariffs vary among four large geographical areas. For solar power, feed-in tariffs were set up in 2011. Subsidies from national bodies are transferred to grid companies, which in turn then buy energy produced by renewable power plants at the price fixed by feed-in tariffs. However, the grid sector is also dominated by monopolistic state-owned companies: the State Grid Corporation of China (SGCC), which operates 80 per cent of the national grid, and the China Southern Grid (CSG), operating in southern regions.

Since 2002, repeated attempts to break the monopoly of the two giants – in particular through the separation of transmission and distribution functions decided by the NDRC – did not meet any success. These attempts also took place under violent forms, with local distribution companies trying to strive to defend their prerogatives. An article in the Chinese review 'New Century' ('*Xin Shiji*') depicts how local distribution companies engaged in a war with national transmission and distribution giants, despite being poorly armed against them (Pu and Qu 2012). The article relates the fights between employees from the SGCC and employees from local distribution companies in Shaanxi and Shandong, that occurred in April 2012. In Shaanxi, more than 200 employees actually took part in the conflict and almost 70 police officers were called in to intervene. The subject matter of the dispute was a transmission 220-kV electricity line, illegally built by the Shaanxi electricity company. According to regulations, the construction of transmission lines of more than 110 kV has to be taken care of by the SGCC. However, this rule aggravates the unfairness of competition between small local distribution companies and national transmission and distribution giants: through high-voltage lines, the latter can import energy from highly productive areas, where purchase prices fixed by the NDRC are lower. In Shaanxi, the article says that the conflict took place between employees from the local distribution company, which was prevented from building a 220-kV electricity line by National Grid employees who were sent to 'protect the lines'.

For over 10 years, the government has been trying to encourage the establishment of market mechanisms in the electricity grid sector. Such a marketisation would theoretically have the effect of breaking the monopolistic situation of the grid giants, of encouraging competition among enterprises of the sector and, in the end, of improving efficiency. However, pushing for the development of local distribution companies would have implied a downgrading of the situation of the two state-owned enterprises, which are both strongly linked to central bodies. Giving full rein to the development of local stakeholders would have also meant exacerbating political decentralisation, as local governments would have jumped at the opportunity of pushing their local champions to compete with the central state's big companies. It would also be difficult to believe that the SGCC, which is today among the world's top ten enterprises in terms of assets, would let this happen without trying to activate its 'personal' networks (among others, its strong links to the Party) and financial power (2.12 trillion yuan of total assets were reported at the end of 2010 (Lewis 2011)). Last but not least, the dominance of market mechanisms for setting up energy prices might also cause a rise in electricity prices, which would then be passed on to consumers – another risk for social stability for a country in which a large majority of people still need subventions in the supply of energy and other services.

To conclude, the dominance of central state actors in the transmission and distribution sector, combined with the State's approval mechanisms for subsidies – still necessary in order to have access to feed-in tariffs and make renewable energy projects profitable – allowed for the central state to keep strict control over the developments of the energy sector.

Table 9.1 Top 15 ranking of wind power enterprises

	Enterprise	Type	Installed Capacity (MW)	Market share (%)
1	Guodian Group	Central SOE	8.941	20
2	Huaneng Group	Central SOE	6.331	14.2
3	Datang Group	Central SOE	5.620	12.6
4	Huadian Group	Central SOE	2.557	5.7
5	China Guangdong Nuclear Power Group	Central SOE	2.365	5.3
6	Guohua	Central SOE	2.346	5.2
7	China Power Investment Group	Central SOE	1.708	3.8
8	Power Beijing	Prov. SOE	1.314	2.9
9	China Resources Co.	Central SOE	977	2.2
10	Suntien Green Energy	Prov. SOE	935	2.1
11	China Energy Conservation and Environmental Protection Group	Central SOE	805	1.8
12	China Wind Power	Private	767	1.7
13	Ningxia Power	Prov. SOE	745	1.7
14	Three Gorges	Central SOE	731	1.6
15	Tianrun	Private	726	1.6
	Others		7.867	17.6

Source: Adapted from Fulton 2012.

Central government control over the investments in renewable energy production and in managing decentralised production projects

Even if Beijing was careful enough to set up safety margins to guard against a takeover by local stakeholders, the wind power sector nevertheless remains dominated by central state-owned enterprises, just as traditional energy sectors – the only difference being the number of subsidiaries. According to the top 15 ranking (see Table 9.1), central SOEs would own at least 70 per cent of market shares.

Moreover, in August 2011, the national government banned provinces from authorising projects of less than 50 MW in size (China's National Energy Administration 2011). This decision was supposedly aimed at addressing concerns whereby provincial governments had been too hasty in encouraging wind power development, which was creating capacity that was under-utilised. The recent grip of the government over the uncontrolled development of wind power capacity is indeed partly motivated by its willingness to encourage enterprises to enter a 'new area of quality'.[11] The wind energy sector was indeed facing several issues. First of all, in the course of its rapid development, several technical problems occurred: among others, the disconnection of installed capacities from the grid as well as curtailment issues (although wind is available, physical and non-physical issues prevent energy from being dispatched into the grids). Incentive policies were targeting installed capacities without taking account of the effective connection of capacities to the grid. In 2011, when the first political measures were taken so as to slow down the pace of development in order to improve the quality of the sector, it was estimated that only one third of wind turbines were connected to the grid. It was also discovered that, in wind farms connected to the grid, local grids were sometimes not able to absorb power generated by new installed capacities.[12] Moreover, the important pace of development did not allow for stakeholders to take the time to answer the multiple technical challenges posed by wind power, like alternative production and overcapacities, which led to damages of grids and power shortages that affected both the enterprises of the sector and the end consumers. Last but not least, the rapid multiplication of enterprises caused a growing competition that drove them to a race of the lowest prices, which seriously altered both the quality of projects and the technological levels of products.[13]

After a period of rapid development, the government thus considerably slowed down the pace of wind energy development. The approval of projects was submitted to new criteria regarding quality, security, industrial standards and connection to the grid, which considerably reduced the number of developments in the sector. In 2011, only 7.6 gigawatt were installed, compared to 18.9 gigawatt in 2010. In 2012, less than 20 gigawatt were authorised by the government, compared to 26.8 gigawatt in 2011. On the other hand, the share of connected wind farms has been significantly rising since 2011: China was able to connect 16 gigawatt to the grid, compared to 13 gigawatt in 2010.[14]

The combination of a dominance of the wind sector by central state-owned enterprises and a set of rules giving central state full control over the approval of new projects has proven particularly efficient in keeping the sector in the hand of the state's planning. The same cannot be said for solar power, however, where the situation is quite different: the sector is mainly run by private stakeholders, as the following Table 9.2 shows

In terms of the development of capacities, the situation of solar power is also different. The growth rate of solar power capacities has been rapidly rising since 2011. It is worth mentioning, however, that solar power installed capacities were far behind those of wind power: in 2010, only 1 gigawatt of solar power was installed, compared to 42 gigawatt of wind power. It was also mainly constituted of water-heaters. Today, however, the government wishes to turn solar power into a new pillar of China's renewable energy strategy. In 2011, installed capacities tripled to reach 3 gigawatt, and they should reach 15 gigawatt by 2015.

Three causal factors explain this turning point in solar development. The first is that in 2011, the government set up a clear grid of feed-in tariffs for

Table 9.2 Top 10 ranking of Chinese solar power enterprises

	Enterprise	Type	Market share (%)
1	Yingli Solar	Private	8
2	Suntech	State-owned (since March 2013[15])	6
3	Trina Solar	Private	6
4	Canadian Solar	Private	5
5	Rene Sola	Private	4
6	Jinko Solar	Private	4
7	JA Solar	Private	3
8	Hanwha SolarOne	Private	3
9	Zhongli Talesun Solar	Private	3
10	Changzhou EGing Photovoltaic Technology	State-owned	2

Source: Adapted from ENF Solar 2013.

solar power. Secondly, the development of new technologies due to advanced research and the increase in their production, which led to a drop in production costs. Finally, the impossibility to keep on relying on international markets for solar panels to be sold – in 2011, as a consequence of the economic crisis hitting former importing countries, international demand dramatically dropped for Chinese PV[16] – has been forcing Chinese enterprises to reorient their efforts towards the domestic market. The approval of solar power projects also proved to be less painful than for wind power projects, as administrative constraints are smaller. Moreover, the use of solar power projects that need less equipment and technical issues regarding installation is less complex: for example, contrary to wind turbines, which are oversized for traditional transportation means, solar panels are easier to transport.

In order to explain this relative 'freedom' of the solar energy sector compared to that of wind energy, it seems that solar is not big enough to pose a serious threat to the government's energy supply management. Moreover, faced with rising issues of air pollution and rising demand, the State is constrained to encourage the development of all forms of solar power capacities. For instance, in July 2013, the Ministry of Finance announced that subsidies would be given to small-scale distributed solar power plants. However, subsidies would be provided indirectly through local electricity distribution companies. Furthermore, as far as energy transition would foresee, China is still quite far from a model in which power is generated by individual households and where their surplus of energy is fed back to the grid.

To conclude, although the situation is quite different from that of the wind sector, the State maintains control tools in the solar sector for the development of enterprises through subsidy mechanisms. However, the scale of the sector being smaller, it poses a minor threat to the traditional model, which offers the solar sector more leeway to develop.

Unsuccessful price reforms help NOCs continue to be the main players

As mentioned above, solar and wind powers would probably not be able to survive without governmental subsidies. This is partly due to the fact that electricity prices are still established by the government. Indeed, technology-intensive wind and solar power sectors would not be profitable without feed-in tariffs given the low electricity prices fixed by the National Energy Administration in order to protect consumers. However, a reform of electricity and the energy prices mechanism is highly recommended for China to respond to its energy security issues and to promote the development of cleaner sources of energy. In terms of questioning 'what should be protected' and 'from what', it is clear that the development of renewable sources of energy would be able to answer the problem of energy shortage by adding another promising source of energy, one that would be capable of being decentralised in its management, as well as be clean for the environment. An important adjustment to energy prices would thus include the costs of the use of coal, a move that could make the use of renewable energies not only more beneficial in environmental terms, but also more appealing in economic terms.

The establishment of prices through market mechanisms would allow for the development of competition between different energy sources. Considering the abundant and cheap coal resources, it is not obvious that such a competition would help develop alternative resources such as renewables. However, it is worth recalling that coal resources are located in the 'far west' of China, whereas consumption centres are mainly in the eastern part of the country. Just by adding transportation costs and the rise in coal prices could incite enterprises to consider the question of choosing among different energy sources. Indeed, if environmental and health costs are then taken into account in the prices of coal, the advantages of switching to cleaner sources of energy could become even more obvious.

However, the most important aspect of a reform of electricity prices could be the possibility to break the NOCs' monopoly. The economic situation of the NOCs has indeed been consistently degrading in the past few years, and this is partially due to the artificial maintenance of their monopoly over power resources, sometimes going against a discourse of profitability for these very same companies. In 2012, PetroChina's (CNPC subsidiary) debt reached 450 billion RMB, and its debt to equity ratio was almost 40 per cent, whereas many private stakeholders in the energy sector were performing much better.[17] Letting market demand and supply regulate prices for energy and electricity could end preferential policies for NOCs and break their monopoly.

The reform of electricity prices has been underway for already a decade. Nonetheless, the government is still reluctant to lead it to its completion because market-based prices would cause an uncontrolled rise in electricity prices – at least in the first years following the reform – which would not be bearable for a large percentage of the Chinese population. This argument, surely partially true, can also be seen as being regularly provided by the government as 'a justification' for the delays of the price reform. However the potential reactions by the Chinese population are not the sole reasons for the unsuccessful price reforms. Market-based prices would probably not allow clean energy sources to compete with coal. However, fixing a price for the coal industry's externalities could be a way to solve this issue. Reforming prices also implies negotiating with all of the sector's stakeholders, perhaps the biggest stumbling block on the path to market prices. Apart from the NOCs and grid giants who have the power to considerably slow down the reform, a multiplicity of stakeholders also wishes to defend their interests. Consumers and the variety of their situations have to be taken into account, from the rich coastal provinces to the most remote and poorest areas in the West of China, from industries to households. Against this background, people's dissatisfaction would then be the last of the problems.

Quasi-consensual forms of decentralised energy

Small hydropower: the real decentralised energy sources

The development of small hydropower has been necessary to provide electricity to remote rural areas. In a context where filling rural–urban gaps (not only in terms of economic development, but also in terms of living conditions and infrastructures) has become a government priority, improving the electricity supply to rural areas has turned into an important task. However, contrary to wind and solar powers, small hydropower is not a new source of energy, for it started as early as the 1950s. By the end of 1995, 19 gigawatt of small hydropower facilities were already installed in rural China. Moreover, in order to encourage the development of small hydropower-based rural electrification, the government authorised the development of local networks of suppliers. Special loans and tax exemptions were also provided for hydropower projects and existing stations.

Although the government established special agencies to provide guidance and coordination for the planning, construction and management of rural electrification projects, the Chinese system's approach of small hydropower is highly decentralised. As Tong Jiandong (2009: 11) points out:

> Except the main policy and strategy, as well as national overall planning which have always been stipulated by the central government, all other aspects such as project planning, project development, implementation and management, operation and maintenance are decentralized to different government levels; i.e., province, prefecture, county or even township.

This policy of 'self-construction, self-management and self-consumption' was set up in the 1960s. From that time, local bodies – whether public, private or collectively owned – were encouraged to invest in the construction of small hydropower infrastructures which they will then manage in order to meet local demand. In a certain sense, it could be seen as a preliminary form of 'energy transition'. This form of decentralised policy made it possible for about 80 per cent of generated electricity power to be distributed to consumers through county and prefecture grids (Tong 2009: 13).

However, as small hydropower remains at a very small scale and intends to feed rural remote areas where traditional energy channels would not be profitable, it is unlikely to pose any threat to energy stakeholders in power. Moreover, small hydropower is just one of the two legs of China's hydro-power strategy – the other one being dominated by large-scale hydro plants (the largest (worldwide) being the well-known Three Gorges Dam in Hubei (22 500 MW)), owned by state actors (the Three Gorges Dam, for example, is operated by a subsidiary of China Three Gorges Corporation (CTGC), a central SOE administered by the SASAC).

Biomass

The government also recently decided to encourage the development of bio-mass. It is important here to notice that all forms of biomass were not equally pushed. In this chapter, we chose to focus on biofuels and biogas in parti-cular, as the two sectors are at both ends of the government's incentive spectrum for biomass.

In the beginning, biofuels represented a real hope for China. It was indeed providing an answer to the country's rising dependency on oil imports. In 2002, five cities (Harbin, Zhengzhou, Nanyang, Luoyang and Zhaodong) were chosen in order to conduct pilot projects intended to develop ethanol vehicles. In 2004, the NDRC decided to extend the experiment to other areas and to set up demonstration sites. In parallel to the development of demon-stration projects, production was growing rapidly. However, in 2006, the gov-ernment started realising that the development of biofuel production could have negative effects on food security. China has to feed 20 per cent of the global population with only 7 per cent of the world's arable land. Moreover, accelerated and uncontrolled urbanisation is impacting the amount and quality of available land for agriculture. Finally, the urban population is rapidly increasing and food diet is quickly evolving towards regimes richer in protein, increasing meat consumption. China managed to replace backyard farming with industrialised livestock farming, but the country still needs to feed these farms (without grazing areas) with land-intensive products such as maize or soybeans. China's grain production cannot currently keep up with the pace of the growth in grain demand. In 2010, the country, which was already relying on imports for soybean, started increasing its corn imports.[18] The conflict with biofuels is thus quite obvious. In 2006, the government decided to strengthen entry regulation. From zero in 2002, corn and wheat ethanol rose to almost 1.5 million tons in 2006, but then quasi-stagnated. Since 2006, the focus has been on non-grain fuel ethanol, but the current technological level and profitability of projects do not encourage enterprises to join the sector. Certain private enterprises have invested in the production of biofuels. However, the sector is currently characterised by a domination of the state-owned enterprises that are also involved in the sector, such as the three oil giants (CNPC, SINOPEC and CNOOC) and the China National Cereals, Oil & Foodstuffs Corp (COFCO). However, compared to other countries, the production remains minor (France produces 1.7 million tons for a population 20 times smaller) and with few prospects for growth.

The situation is quite different for biogas. In the second half of the years 2000, the government deployed a set of subsidies to help the development of digesters. Between 2005 and 2011, the government spent 21.2 billion RMB towards the development of the sector, with a focus on less developed areas such as Western China and on grain growing areas in the centre and in the north-west of the country. However, the development of capacities has proven to face some obstacles: for example, in the north of the country, low winter temperatures hinder the continuation of gases fermentation and call for specific systems and technologies (isolation, heating system) Moreover, many projects that involve small farmers (there were 41.68 million household small digesters in rural areas in 2012) are currently being abandoned. Economic opportunities in cities are emptying the countryside of trained young people, and elderly farmers who remain in rural areas are not physically able to maintain biogas equipment. Moreover, biogas did not necessarily encounter positive user feedback.[19]

China also tried to raise interest in biogas in big industrialised livestock farms that have been mushrooming in the east of the country for some years. In 2012, there were more than 80 000 units of biogas plants in livestock farms. However, digesters are still frequently seen as a way of drying manure in order to produce fertiliser that can then be sold to farmers. Electricity is seen as a secondary product of digesters and it is not always even used by the farm (the connection to the grid also remains very rare).

These two examples – biofuels and biogas – highlight the fact that biomass sectors are treated very differently by the government and do not face the same problems on their path of development. Nevertheless, the fact that biomass is competing with cheap sources of electricity does not encourage actors to join the sector. This alternative source of energy remains on a small-scale and thus is unable to pose a threat to existing powerful stakeholders.

Conclusion

In the framework of its twofold energy strategy (diversifying resources and reducing the share of coal in the energy mix), China has been actively pushing for the development of alternative sources of energy. Renewables have been taking an important place in this strategy. These also offer the possibility for local stakeholders to build distributed generation plants. This model is very different from the traditional centralised energy production model, in which big national stakeholders, under the control of the central state, run energy production at a national level. This article has shown that the development of decentralised energy sources is still strongly regulated by the government. Today, most of the alternative sources analysed in this article are still unable to compete with traditional energy sources, which still fundamentally hinder their development. These sources must strongly rely on governmental subsidies, yet through the authority held over subsidies for wind and solar power, the government is able to remain in control and limit the potential that could significantly trigger the basis for an energy transition. Moreover, certain subsidies that are necessary to the daily survival of wind and solar power enterprises go through state-owned grid companies, reducing the possibilities for other actors to enter this promising market. Additionally, for the wind sector, enterprises are mainly owned by the State. Finally, the government remains in charge of setting prices for electricity, most often in favour of already powerful stakeholders (the NOCs and grid giants). Repeated attempts, stretching over more than a decade, to reform energy prices have not yet produced any satisfactory results.

In a nutshell, the development of alternative resources of energy, though highly desired by the central state, also highlights conflicts between state-owned enterprises and smaller local players. Moreover, the question of power allocation between the central state and local governments has never been so strong than in the energy sector, particularly with the development of decentralised energy. However, the central government still manages to control stakeholders through mechanisms and institutions. The example of small hydropower, however, has proven that a transition towards mixed forms of decentralised and centralised energy production centres remains possible in China and indeed that it could be the answer to the country's energy security problem.

Notes

1 Many political scientists saw in the country's rapid economic development the reasons for the population's acceptance of an authoritarian regime. The association between growth and social and political stability is also deeply engraved in the Chinese government's mind.
2 'The power and autonomy of China's NOCs is due to a number of factors, including their relative strength vis-à-vis the central government's energy bureaucracy' (Downs 2010: 76).
3 GDF (Gaz de France), EDF (Electricité de France) and Charbonnages de France were created by the law of April 8, 1946, which nationalised electricity and gas in France (EDF and GDF will be listed on the stock exchange in 2005). In 1947, the United Kingdom nationalised public and private actors involved in the production and transmission of electricity. Endesa was created in Spain in 1944 (before being privatised in 1988).
4 Muller-Kraenner talks about countries entering a 'New Great Game' (Müller-Kraenner 2008).
5 'Increase in China's energy consumption between 2000 and 2008 was more than four times greater than in the previous decade.' (IEA 2010: 5).

6 In 2007, the World Bank estimated that the economic loss due to pollution was reaching almost 6 per cent of GDP (World Bank 2007).
7 The World Health Organization's recommended standard is 25.
8 In the past, China's energy strategy essentially focused on diversification (both geographical and per type of resource) and securing resources (among others through raising the technological level of the sector).
9 Targets initially set by the 12th Five-Year Plan.
10 Polysilicon is used to manufacture solar photovoltaic cells and panels.
11 Asia Centre Energy round tables, June 2012.
12 Electricity grids are designed to absorb designated quantities of energy. The limited voltage of lines designed to absorb a certain quantity of energy will prevent grids from coping with additional volumes.
13 Asia Centre Energy round tables, June 2012.
14 Asia Centre Energy round tables, June 2011.
15 In 2012–2013, the drop in foreign demand led to an important crisis in the Chinese solar sector. Suntech, which was on the verge of bankruptcy, was bought by the state and turned into a state-owned company. But apart from Suntech, which was 'too big to fail', others remained private.
16 In addition, the establishment of tariffs also impacted trade when entering the EU and US markets in March 2012, set up to limit Chinese imports.
17 The rise in PetroChina's debt is partly linked to its operating losses from gas imports: the rise in gas prices on international markets and the growing gap between production and consumption forced the company to import larger volumes at higher prices, whereas in the meantime, energy prices were maintained at a fixed level by authorities.
18 The government is perfectly aware of the fact that relying on imports is risky, not only for domestic price stability but also for other importing countries (with a population of 1.3 billion, China cannot adopt Japan's strategy of relying on imports for food). The country set up a 95 per cent grain self-sufficiency target which it wishes to maintain, in spite of the growth in grain demand and the decrease in available production resources.
19 According to an expert from Planet Finance (interviewed in June 2013), migrants' experiences with 'instant' energy led them to view biogas as primitive and arduous to maintain, and households often switched to LPG and cheap electricity.

Bibliography

China Renewable Energy Industry Association (2012) *Zhongguo fengdian fazhan baogao* [*China Wind Power Outlook*]. Beijing: CREIA.

China's National Energy Administration (2011) *Guojia nengyuanju wenjian guoneng xinneng no. 285* [*New Regulatory Policy on Wind Power no. 285*]. Beijing: China's National Energy Administration.

Dan, S. (2007) 'Structural Reforms in China's Oil Industry: Achievements, Problems, and Measures for Further Reform.' In M. Meidan (ed.), *Shaping China's Energy Security. The Inside Perspective.* Paris: Asia Centre, pp. 113–124.

Downs, E. S. (2007) 'China's Energy Bureaucracy: The Challenge of Getting the Institutions Right.' In M. Meidan (ed.), *Shaping China's Energy Security. The Inside Perspective.* Paris: Asia Centre, pp. 64–89.

Downs, E. S. (2010) 'Who's Afraid of China's National Oil Companies?' In C. Pascual and J. Elking (eds), *Energy Security: Economics, Politics, Strategies and Implications.* Washington, DC: Brookings Institution Press, pp. 73–102.

ENF Solar (2013) *Top 10 Chinese Solar Power Enterprises.* Available at: www.enf.com. cn/top10 (accessed 12 February 2014).

Fulton, M. (ed.) (2012) *Scaling Wind and Solar Power in China: Building the Grid to Meet Targets.* Frankfurt: Deutsche Bank Group.

International Energy Agency (IEA) (2010) *World Energy Outlook 2010 – Executive Summary.* Paris: IEA.

International Energy Agency (2011) *World Energy Outlook.* Paris: IEA.

Lewis, C. (2011) 'China Power Reform Spinoff Costs SGCC 4 pct of Assets', *Reuters,* 3 November. Available at: www.reuters.com/article/2011/11/03/china-power-sgcc -idAFL4E7M31A220111103 (accessed 10 February 2014).

Lieberthal, K. and Oksenberg, M. (1988) *Policy Making in China: Leaders, Structures, and Processes.* Princeton, NJ: Princeton University Press.

Müller-Kraenner, S. (2008) *Energy Security.* London: Earthscan.

Naughton, B. (1995) *Growing Out of the Plan: Chinese Economic Reform, 1978–1993.* Cambridge: Cambridge University Press.

Pu, J. and Qu, Y. X. (2012) 'Dianwang zhanzheng: dianli tizhi juyou zi xia er shang gaige keneng [Grid War: The Electricity Reform Might Come From Below].' *Xin shiji* [*New Century*] (21 May).

Schwoob, M. H. (2011) 'Énergie en Chine: Les Voies Engourdies du Charbon.' *China Analysis* 35: 42–44.

Tong, J. (2009) 'Some Features of China's Small Hydropower.' *Hydro Nepal* 4: 11–14.

The World Bank (2007) *Cost of Pollution in China: Economic Estimates of Physical Damages.* Washington, DC: The World Bank.

Waldron, S., Brown, C. and Longworth, J. (2006) 'State Sector Reform and Agriculture in China.' *The China Quarterly* 186: 277–294.

10 A pillar of national energy security

Industrial energy efficiency

Patrick Schroeder

Introduction

At the most general level, the term energy security can best be understood as robustness against disruptions, including sudden and unforeseen interruptions, to energy supply (Moomaw *et al.* 2011). In practice, energy security means different things in different countries, largely depending on their stage of development. From the traditional perspective of industrialised countries, it means improving the means to secure low-cost and reliable supplies of fossil fuels for electricity generation and transport. For many developing countries, energy security is mostly about meeting basic human needs at the household level, as per capita consumption levels and the quality of energy supplies are often far lower than in OECD countries (GNESD 2010). Increasingly, energy security is considered an issue that is inseparable from low-carbon development and climate protection, and renewable energy options can contribute to reaching energy security goals through diversifying energy supplies and diminishing dependence on limited suppliers (Moomaw *et al.* 2011). All of these aspects are highly relevant for China's energy security situation. Furthermore, energy efficiency is a major crosscutting element of China's energy security strategy, as the country seeks to manage growth in demand in order to reduce total energy consumption and thus achieve energy security both nationally and locally – or at least alleviate the current state of energy insecurity.

The most pressing issue that demonstrates China's current energy security concerns power shortages and blackouts. From the late 1990s through to the middle of the first decade of the new century, 26 of the 30 Chinese provinces experienced blackouts associated with electricity shortages, driven by rapid economic growth that outpaced power capacity (Fisher-Vanden *et al.* 2010). China has once again been experiencing shortages since 2008, which became severe in 2011 in at least 12 provinces where the shortfall was at least 2,000 MW (Yu *et al.* 2013). In contrast to earlier power shortages in the 1980s and early 2000s, Yu *et al.* (2013) argue that China now has sufficient power generation capacity to meet the demand, or at least could have if institutional arrangements, particularly market pricing mechanisms, were more favourable to efficient generation and demand-side energy efficiency. According to Wu *et al.* (2012), energy efficiency improvements in China's industrial sector have so far been driven mainly by capital-intensive technological improvements. In an empirical study of 28 provinces in China, Wu *et al.* (2012) also demonstrate that there is further potential for efficiency, as China's provinces are capable of reducing energy consumption by 18.4 per cent annually through energy efficiency improvements, particularly in industry.

Further progress would require the active involvement of stakeholders, ranging from industry right down to individual households. A particular challenge in all countries, which also holds true for China, is the active engagement of small and medium-sized enterprises (SMEs) as end consumers. This challenge goes beyond mere technological efficiency improvements. One major barrier is the lack of sufficient knowledge among SMEs about the possibilities and potential of energy efficiency. In order to overcome this, introducing measures to improve communication is therefore extremely important, but doing so has proven to be decidedly difficult (Lundqvist and Mattsson 2011).

Overall, given the current state of China's development, addressing energy insecurity is a more complex issue for it than for many other countries. Meeting growing demand cannot follow the current approach of increasing the capacity for coal power generation, as pollution levels and CO_2 emissions have already exceeded sustainable limits and are posing a threat to the resilience of China's development. Further, solving China's domestic energy shortage cannot rely solely on an energy demand policy in the narrow sense, as has been traditionally the case in China. According to Zhang (2011), energy security for China will require combining energy policy with macro-economic policy – such as fiscal and monetary policies – and with foreign policy and international cooperation. This is also true for industrial energy efficiency as a crucial pillar of energy security. For this reason, after an overview of China's energy consumption trends and policies for energy efficiency, the chapter will focus on the international cooperation aspect of China's energy efficiency. In particular, it will discuss programmes for policy transfer and technical support for SMEs that manufacture energy efficiency equipment. We will also present and analyse some case studies concerning energy efficiency and energy security in the context of EU–China cooperation under the SWITCH-Asia Programme.

Energy consumption trends in China

One of the most striking features of China's energy sector is the continued growth in energy consumption. Since 1980 China's total primary energy consumption has increased more than fivefold, from 17.2874 quadrillion British thermal unit (btu) to 100.8814 quadrillion btu in 2010 (US EIA 2014). The growth is even higher in electricity consumption, having increased 13-fold from 261.5 billion kWh in 1980 to 3,633.8 billion kWh in 2010 (ibid.). In addition to efforts to meet the growing demand, striving to slow the growth in power consumption is a key element in China's energy security strategy, although results have so far been suboptimal.

China's Bureau of Statistics reported in early 2013 that the country's energy consumption in 2012 was 3.62 billion tonnes of coal equivalent (tce). According to preliminary statistics, China's overall energy consumption rose by 3.9 per cent in 2013 to 3.76 billion tce. Table 10.1 provides a breakdown of overall power consumption and some of the growth trends in the electricity consumption of various sectors in China. It shows that heavy secondary industry is still the major consumer of power in China, accounting for 60 per cent of total power consumption, with a continued strong growth in 2013.

In looking at the largest energy consumer, the industrial sector, we see that total energy consumption continued to rise over the course of the 11th Five-year Plan. The secondary heavy industry sector, in particular, accounts for the lion's share of energy consumption. Coal, meanwhile, remains the leading source of energy, both for electricity generation and heating, representing 77 per cent of production and 68 per cent of energy consumption in 2010. It is anticipated that it will still account for about 60 per cent of energy needs by 2030, according to BP (2012).

There are, however, large regional differences in industrial energy efficiency. According to research by Wang *et al.* (2012), during the period 2005–2009, the areas where energy efficiency of industrial sectors attained the highest levels of efficiency were mainly in the coastal provinces and municipalities of the east, such as Tianjin, Shanghai, Jiangsu, Shandong, Guangdong. Energy efficiency of the industrial sectors of the central and western regions is still relatively low and in some cases up to 40 per cent or more below that of the

Table 10.1 Breakdown of China's power consumption Jan.–Nov. 2013

	Nov. 2013 (bln kWh)	Y/Y pct change	Jan.–Nov. 2013 (bln kWh)	Y/Y pct chg
Total power use	448.5	8.5	4,831.0	7.5
Of which				
Residential	48.7	7.2	625.4	9.4
Non-residential	399.8		4,205.7	
Primary industry	7.8	7.0	93.5	0.1
Secondary industry	343.7	8.6	3,539.1	6.9
Of which				
Light industry	54.5	6.5	580.8	6.2
Heavy industry	283.2	9.0	2,897.8	6.9
Service industry	48.3	8.8	573.1	10.8

Source: Reuters (2013), citing data from China's National Energy Administration. Available at: www.nea.gov.cn.

developed eastern provinces. For instance, if, for argument's sake, the total factor energy efficiency for Tianjin is 1, then the total factor energy efficiency for Shanxi and Hubei provinces is 0.5 and 0.66 respectively (Wang *et al.* 2012). The reason for this is that industrial sectors of these areas are mostly still at the stage of relying on the large input scale of extensive resources, meaning that improvements in technology are urgently needed.

China's legislation and policies for energy efficiency

China's legislation and policies for energy efficiency have been in place ever since the opening-up period initiated by Deng Xiaoping in 1980 through a strategy of reducing energy intensity. The goal was to quadruple GDP while only doubling energy consumption over a 20-year period (1980–2000). China surpassed this goal both in the increase of its GDP, which grew more than five-fold, and a reduction of energy intensity (Levine *et al.* 2009). These energy efficiency measures were designed to reduce the power shortages that plagued many industrial facilities, where interruptions to power meant that normal production was reduced to two or three days a week (Yu *et al.* 2013). A specialised national institution, the Bureau of Energy-Saving and Comprehensive Energy Utilization in the State Planning Commission (SPC), the forerunner of the National Development and Reform Commission (NDRC), was responsible for overseeing and guiding the work on energy efficiency through establishing and operating about 200 local energy conservation centres.

By the end of the twentieth century, China began paying less attention to energy efficiency, as generation capacity had increased significantly and power shortages became less frequent. The governance system for energy efficiency was neglected, which resulted in dramatic increases in energy consumption from 2002 onwards. The reason for this sudden increase after two decades of successful improvements was thus a governance issue, not a technical one. For instance, at the industrial-enterprise level, even though, by law, all key industries were required to have an energy manager, many large enterprises had energy managers in name only, if at all. In fact, there was a reduction in public budgets for energy efficiency and a lack of authority or willingness in national, provincial, municipal and local government bodies to enforce laws and regulations that were intended to save energy (Levine *et al.* 2009). By 2004, senior Chinese government officials had recognised that the rapid increase in energy demand presented serious problems for balanced economic development and, in addition, that it was creating environmental pollution. Accordingly, during 2004 government agencies and think tanks across China were engaged in a re-evaluation of China's energy policy. This led to the 2004 Medium and Long-Term Energy Conservation Plan (National Development and Reform Commission 2004), which set a mandatory 20 per cent reduction in energy intensity by 2010, an annual average of 3.6 per cent per annum, and determined that this downward trend was to continue at the same rate until

2020 (Andrews-Speed 2009). China has almost reached its target, having achieved a 19.1 per cent reduction in energy intensity during this period (Qiu 2011).

A wide range of additional policy instruments were introduced by the Chinese government to support the implementation of effective demand-side management, including sector-specific energy consumption reduction targets, taxation, flexible financing and market mechanisms, technical training, energy performance standards and the labelling of appliances and awareness raising campaigns were being implemented by the Chinese government. Following these changes, China's Energy Conservation Law (National Government of the People's Republic of China 2007) was revised on October 28, 2007, and the new law entered into force as of April 1, 2008. Regarding industrial energy conservation and efficiency, which is the major focus, the revised law includes most industry sectors and requires formulating energy conservation technological policies for electric power, iron and steel, non-ferrous metals, building materials, oil processing, chemical, coal and other main energy-consuming industries (Article 30). Furthermore, it encourages industrial enterprises to adopt efficient and energy-saving motors, boilers, kilns, fans, pumps and other equipment, as well as the technologies of cogeneration of heat and power, waste heat and pressure generation, clean coal and advanced energy consumption monitoring and control (Article 31). Beyond the industrial sectors, the law further requires construction projects, the transport sector and government buildings to cut their energy consumption. It specifies energy management systems as well as a system of rewards and punishment rules with regard to energy conservation. Finally, it set up a target responsibility and evaluation system to save energy and requires that work carried out by local government officials in energy conservation should be integrated into the assessment of their political performance.

After these various policy developments, the current Chinese institutional framework for energy efficiency can be described as a combination of mandatory policies and incentive policies (see Fig. 10.1).

At the time of writing, the 'Detailed working plan for the 12th Five-year Plan on energy conservation and emissions reductions', issued by the State Council, is the main policy document for the years 2011–2015, with the purpose of outlining specific actions for energy conservation in this five-year period. It also sets two energy conservation targets for this period: first, a 16 per cent reduction in energy consumption per unit of GDP compared with 2010, representing a 32 per cent reduction compared with 2005; second, an energy saving target of 6.7 million tce ($\approx$4.7 million tonnes of oil equivalent). It is expected that most of the energy conservation will be achieved through industrial energy efficiency measures. The document also specifically encourages the use of market mechanisms for energy conservation to allow for more flexibility. However, no guiding policy document yet exists at a national level to provide greater clarity as to which types of market instruments are to be implemented. The use of market mechanisms shows some

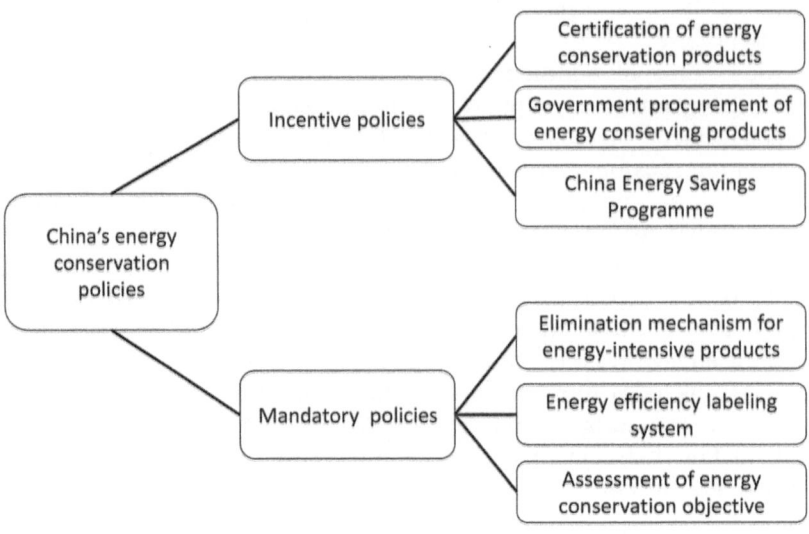

Figure 10.1 China's institutional framework for energy efficiency (after Cai 2013)

degree of policy innovation, given that the Chinese government tradition-ally relied in the main on top-down command and control measures, such as industrial shutdowns, as occurred in 2010, when steel mills in Hebei, Shandong and Henan provinces were ordered to cut output from Septem-ber onwards of that year as Beijing was under pressure to meet the energy efficiency target stipulated in the 11th Five-year plan (Lian and Serapio 2013). Whilst such measures are not in line with energy security objectives that aim to secure a reliable supply to industry stakeholders, further such measures can be expected towards the end of the 12th Five-Year Plan if energy efficiency targets have not been met. Moreover, there are increasing synergies between energy efficiency policies and those aiming for the reduc-tion of industrial air pollution. A shutdown of highly polluting industrial facilities contributing to smog in Chinese cities would also address energy security concerns, as these facilities are often extremely inefficient. Hebei Province will be a particular target of such measures. The province's four major heavy-polluting industries are iron, cement, electricity and glass. Accord-ing to government plans, 94 obsolete iron and steel sintering plants will shut down, iron and steel production will be reduced by 60 million tons, and coal consumption by 40 million tons by 2017 (Zheng 2013).

The most recent policy development has been, at the time of writing, the approval of an 'energy consumption control target' by China's State Council in late January 2013 to keep the country's total energy consumption below the equivalent of 4 billion tonnes of coal per year by 2015, meaning the Chinese government is trying to cap total energy growth around 4.3 per cent until 2015, compared to the 6.6 per cent annual increase between 2006 and 2010 (Xinhua 2013). Although this is mainly aimed at controlling air pollution and CO_2 emissions, it can also be seen as an energy security target. This is not the first time the national government has adopted a 'control target' for total energy consumption. The 11th Five-Year Plan for Energy Development for instance set the 'control target for total energy consumption' at approximately 2.7 billion TCE. However, total energy consumption in 2010 was 3.25 billion tce, or 20 per cent higher than the target (Song 2013), showing that objectives are not necessarily binding and not always effective.

Voluntary Agreements: the 'Top 1,000 Enterprises' and 'Top 10,000 Enterprises' programmes

To achieve the economy-wide energy efficiency targets outlined above, engagement and buy-in of industry stakeholders is crucial. The 'Top 1,000 Enterprises Programme' launched by the NDRC in 2006, was intended to engage major industry players through a voluntary agreement (also called negotiated agreement), an instrument for improving environmental performance of industry, including energy efficiency, that has been adopted and practised in the EU, Japan, the USA and some other countries for decades (EEA 1997). It is not always possible to ensure a one-to-one transfer to developing countries of instruments that have been successful in industrialised countries. To experiment with the transfer of this instrument to China, the European Union (EU) funded a demonstration project that was implemented in two stages between 2004 and 2009 under the Asia-Pro Eco II Programme. In the first phase, 29 companies were involved, while the second phase saw the participation of 14 companies from Nanjing, Xi'an and Kelamayi, and voluntary agreements were signed between the three local Environmental Protection Bureaus and state-owned companies from different energy-intensive sectors (steel, power generation, cement, refineries, breweries, petrochemical). The pilot agreements were informed by Dutch experiences with long-term agreements on energy efficiency, but required significant adjustment to the Chinese circumstances (Eichhorst and Bongardt 2009).

The Top 1,000 Enterprises Programme was China's first nationwide voluntary agreement. Accordingly, it covered the 1,008 largest enterprises whose energy consumption exceeded 180,000 tons of coal equivalent (tce) ($\approx$125,500 tons of oil equivalent or toe), representing about one third of China's national industrial energy consumption at that time. This programme was one of the key initiatives for realising the above-mentioned goal of the 11th Five-year Plan (2006–2010) for a 20 per cent energy intensity reduction. The National Development and Reform Commission drew up the list of enterprises jointly with provincial governments, which then set the targets for each enterprise.

Although the programme was promoted as a 'voluntary agreement' between the government and industry, the target-setting process was carried out in a top-down manner with targets determined by the National Development and Reform Commission (NDRC). The concrete implementation for the five-year period was delegated to the provincial and municipal levels. All participating enterprises signed energy conservation agreements with local governments and promised to reach the energy savings target in the five-year period. In the case of Beijing, the NDRC signed an agreement with the Beijing Municipal Government covering ten enterprises within Beijing's jurisdiction. The Beijing Municipal Government in turn signed energy-efficiency target contracts that included energy saving amounts with each of the ten enterprises (Price *et al.* 2008). The energy-saving authorities of the respective province, district, or city were directed to collaborate with related organisations to lead and implement the Top 1,000 programme, including the tracking, supervision and management of the energy-saving activities of the enterprises. Moreover, the local authorities were directed to oversee and 'encourage' the enterprises in their energy management, energy auditing, and energy reporting requirements (ibid.).

Although the enterprises selected to participate did not actually have a real choice to opt out, the programme was not executed merely in a command and control approach, as many of the large state-owned enterprises are powerful enough to resist such measures. Therefore, the programme also included a number of financial incentives for enterprises. The participating companies were thus awarded financial support at a rate of 200 RMB for every tce saved each year for enterprises in East China, and 250 RMB for every tce saved per year for enterprises in Mid or West China. Furthermore, through the support of the UNDP and the EU–China Energy and Environment Programme, a number of capacity-building training courses on energy auditing and energy management were conducted (Kan 2008). This resulted in the introduction of full-time or part-time energy managers in more than 95 per cent of the Top 1,000 enterprises.

The programme suffered, however, from the lack of a rigorous assessment of potential energy savings, as well as from the fact that it set somewhat arbitrary targets that left many potential efficiencies unexploited or, as happened in some cases, it allocated unrealistic targets to companies (Xue and Bressers 2010). This is related to an on-going problem undermining effective environmental governance, due to the unwillingness of many companies to disclose information about energy use, pollution discharges and emissions. Furthermore, the programme only covered some enterprises in certain sectors, and left a large part of industrial energy consumption unregulated. Despite these shortcomings, the programme is generally considered to be a success (ibid.). The NDRC produced a Top 1,000 programme performance evaluation report by the end of 2009. According to this report, there were in total 922 enterprises evaluated, 886 of which had achieved their targets for the year. Only 36 failed to fulfil the agreed targets. The total energy saved in 2008 was 35.72 million tce. By the end of 2008 the programme had saved 106.20 million tce in total, achieving the target of 100 million tce two years before the 2010 deadline (cited in Xue and Bressers 2010). By the end of 2009, the programme had exceeded the initial target and saved 124 million tce in its first three years (Finnamore and Davidson 2011). The example of the Top 1,000 Enterprise Programme shows that China has quite successfully adopted and adapted the mechanism of voluntary agreements (Xue and Bressers 2010). It also shows that companies can improve their energy efficiency performance, if strong top-down targets are complemented with financial incentives and capacity building.

This positive example eventually led to the creation of a similar follow-up program, the 'Top 10,000 Enterprises Programme'. This new version uses the same principle, but has a lower inclusion threshold, as enterprises are included if their annual energy consumption is over 10,000 tce ($\approx$7,000 toe). Smaller enterprises, with energy consumption of more than 5,000 tce ($\approx$3,500 toe) may also be included by administrative authorities, if they are considered to be 'key units'. The energy consumption of the 'Top 10,000 Enterprises' represents more than 60 per cent of total Chinese energy consumption. In May 2012, the NDRC drew up the list of the 'Top 10,000 Enterprises' and the energy savings target for each province based on regional proposals. The target set is for 15,000 enterprises, including small and medium-sized enterprises, to save 250m tce ($\approx$174m toe) in 2016 compared with 2011. The share of each sector varies among provinces, with the majority taken up by the industrial sector, followed by the transport, services and retail sectors. Equipment upgrades will be one major component to achieve the targets set by the programme. Other key elements include the establishment of energy conservation working groups in enterprises, conducting energy audits and developing energy conservation plans, and implementing energy audit systems, energy efficiency benchmarking and capacity building for energy managers (Reinaud and Goldberg 2012). The programme includes enforcement mechanisms and sanctions. If enterprises do not meet the annual energy-saving targets, mandatory energy audits will be conducted and adjustments/retrofits will be required to be made within a limited period of time. Preliminary results of the activities of the programme were not available at the time of writing.

SWITCH-Asia project case studies: promoting efficient motor systems and transformers in SMEs

International cooperation and the exchange of international best practices for energy efficiency is an important approach for contributing to China's energy security. The SWITCH-Asia Programme, funded by the EU and local match-funds, has been implementing projects in Asia and China for improving energy efficiency in key industry sectors since 2009. The SWITCH-Asia Programme has been established by the EU's Directorate-General for Aid Cooperation and is one of the EU's largest development programmes with

Asia, with the aim of promoting sustainable consumption and production patterns. Although not specifically mentioning energy security as an objective, the programme tries to address the challenges of insufficient technological investments in the energy sector and the failure to reach the scale of manufacture of energy-saving industrial equipment, two major factors preventing China, and other Asian developing countries, from achieving higher energy efficiency (see Wang *et al.* 2012). The particular focus of the programme are SMEs, which face specific challenges such as a low awareness of energy efficiency, lack of management structures for energy efficiency, out-dated and inefficient equipment, and limited access to finance to make investments in energy efficiency. The SWITCH-Asia projects in China therefore complement previous cooperation projects, such as the 'China End Use Energy Efficiency Project', funded by UNDP for a 12-year period, starting in 2001, which mainly focuses on large state-owned enterprises. Although small in size, due to their large number, SMEs account for a significant share of China's energy consumption. The following section gives an overview of the strategies, approaches and challenges of three projects carried out under the SWITCH-Asia programme focusing on two key technologies for China's energy efficiency in industry: electric motor systems and transformers. These technologies are used across many different industry sectors ranging from cement, steel, energy generation to chemicals and many more. Improving the performance of these technologies and replacing out-dated equipment with high efficient motor systems and transformers is therefore crucial to reduce energy consumption in these sectors. Each case study will describe the approaches of each project to achieve technological improvements for higher energy efficiency.[1]

Case study 1: Electric motor systems in industrial production

Electric motor systems are utilised throughout all major industrial sectors where they drive both core industrial processes, like presses or rolls, and auxiliary systems, like compressed air generation, ventilation or water pumping. As a result of their wide application they account for about 60 per cent of global industrial electricity consumption (Fleiter *et al.* 2011). Improving efficiency of these motor systems is therefore key to any energy efficiency and energy security strategy. Many companies are often unaware of the huge potential savings in energy and the quick return on investment for upgraded motor systems, particularly in small and medium-sized enterprises. The challenge is to raise awareness of the true cost of the motor systems and to channel investment into replacing inefficient systems. Electricity consumption by electric motors in China is projected to grow to approximately 3,000 TWh (Terawatt-hour) in 2020, from 1,400 TWh in 2006 (Yang 2007), accounting for about 60 per cent of total electricity consumption in the country. Currently, their actual operational efficiency is about 10–30 per cent below international best practice, depending on the industry sector. Over 70 per cent of the motors in China have a capacity between 10 and 200 kW/unit. As the majority of electricity in China is generated from coal, causing the average amount of CO_2 per kWh to be higher than in developed countries, electric motor systems are a significant contributor to climate change. To give an idea of the energy efficiency potential for motor systems, Yang (2007) estimated that installing intelligent motor controls to large induction motors in China, a reduction of 575 million tons of CO_2, 4.6 million tons of SO_2, and 1.8 million tons of NOx could be achieved between 2007 and 2020.

The SWITCH-Asia project 'China Motor Challenge',[2] implemented from November 2008 until February 2012 by the China National Institute of Standardization (CNIS),[3] aimed to facilitate over 400 major industrial users of electric motor systems to improve the operating efficiency of their systems, and achieve a far-reaching impact in the demand for high-efficiency motor systems, while actively supporting the creation of a stimulating policy environment. This was the most challenging part of the project: the transformation of the market towards high efficiency electric motors. Other objectives were comparatively easy to achieve, such as the promotion of best practice in the design and application of energy-efficient motor systems, the promotion of energy service companies (ESCOs) and their services, and increased exports in Chinese goods that meet international standards.

Through capacity-building workshops, the users of motor systems were not only trained in their efficient use but also in how to upgrade their existing systems. The motor systems users were provided with information about ESCOs active in their area and sector, which can supply know-how, equipment and financing to realise the upgrade. Over the three-year duration of the project around 400 major industrial users were engaged to improve the operating efficiency of their electric motor systems. Three hundred producers of motor systems took part in workshops dealing with international standards and how to achieve and comply with these standards in their companies. This helped them understand how to adapt their products to international requirements and how to prepare for the upcoming new Chinese energy label for electric motors. The project also created an impact by promoting strategic and operational changes in the industry through the development of institutional partnerships among stakeholders in the value chain. Here, the role of energy service companies (ESCOs) is crucial. ESCOs are still a rather new type of service company in China. They typically facilitate 'energy performance contracts' where the service company takes on all upfront costs of an energy audit and subsequent purchase of equipment, and this is paid according to the energy savings achieved. This method effectively reduces the risks for the customer. The project focused on around 80 companies which are active in delivering services that are related to energy savings through motor systems upgrades. They were trained in best practices in the design and application of energy-efficient motor systems and how to expand their business models by adding new services. The project established a 'China Motor Systems Challenge Club', an information platform connecting users and producers of motor systems, energy service companies and public energy administrators. Through this club, best practices in the design and application of energy-efficient motor systems are promoted and shared. An annual award ceremony recognised cases where excellent energy-savings were achieved. It also established a national information platform for Chinese motor system users, energy service companies, electric motor manufacturers and energy-saving administrations, in order to promote motor system upgrades.

By February 2012, more than 1,000 industrial motor system users and 300 energy service companies had taken part in the training workshops. Most of them registered as members of the China Motor Systems Challenge Club and 400 upgraded their motor systems. In addition, almost 300 motor producers were made aware, through these workshops, of the latest development in international standards and their related regulations. Finally, the project actively supported the improvement of a policy environment conducive to promoting industrial energy efficiency. China's national energy efficiency standards for electric motors were outdated at the start of the project, meaning their level of efficiency was much below international levels. The project contributed to the development of new national standards, based on those currently used in the EU. The mechanism introduced will increase minimum efficiency requirements through a minimum energy performance standard (MEPS), the main policy tool for improving efficiency in technical equipment. In terms of policy, motor systems are a good example of how mandatory standards and voluntary instruments can work together with financial incentives. The first Chinese standard on minimum allowable values of energy efficiency and energy efficiency grades for small and medium three-phase asynchronous motors, the most widely used motor type in industrial production, was first introduced in 2002. It was subsequently updated in 2006, and again in 2012, partly based on the recommendations of the project. Complementary measures are the energy efficiency label for motor systems issued in 2004 by the NDRC and the General Administration of Quality Supervision, Inspection and Quarantine (AQSIQ). Another recent development is the subsidy policy for energy efficiency motor system products. In May 2010, the NDRC and the Ministry of Finance (MOF) jointly included high-efficiency motors in the large national financial subsidy programme called the 'China Energy Savings Programme'. Such policy efforts tend to have a lasting impact on the market, as the new standards will remove outmoded, low-efficiency products from the market. A new label will provide clear and simple information to users, enabling them to make better-informed decisions when making purchases. A supervisory mechanism will ensure that the efficiency claims on products are genuine. This cooperation project is not the last word on improving motor system efficiency, however, and the standards which were introduced through the support of the project will need to be revised and improved again in due time. The uptake of efficient motor systems by SMEs is still lagging behind, despite the implementation of the new national standards in 2012, and market transformation will take some time.

Case study 2: Efficient transformers

A transformer is an electrical device that transfers energy between two circuits. A broad range of transformer designs are used in electronic and electric power applications, such as interconnecting the power grid. In China, the annual loss of electricity is estimated to be more than 20 billion kWh, and about 30–40 per cent of this loss derives from power transmission and distribution (SWITCH-Asia Network Facility 2012). Power loss as a consequence of inefficiencies associated with the transformers themselves amounts to more than 3 per cent of the total installed capacity of China's national grid (International Copper Association 2010). Whilst major energy-intensive industries use large quantities of transformers, but upgrading those that are inefficient is not economical. Indeed, due to China's low electricity prices for industry users, which range from 0.3 to 1 RMB per KWh depending on volume and time of consumption, the cost savings generated through reduced energy losses are not substantial enough to compensate for the investment cost. However, energy savings can be substantial, depending on the needs of the distribution network. Indeed, savings of between 10 per cent and 35 per cent are possible when replacing older technology with a high-efficiency transformer. China has around 1000 manufacturers of transformers, most of them SMEs, which are engaged in fierce competition, as the market is suffering from excess capacity and decline in transformer quality (Cai 2013). Furthermore, in many cases local manufacturers lack the capacity and awareness to produce transformers with higher efficiency, whilst end-users do not see the advantages of using them. The SWITCH-Asia project 'China Higher Efficiency Power and Distribution Transformers Promotion',[4] which was implemented between December 2009 and December 2012 by the International Copper Association China, reduced electricity losses by increasing the market penetration of higher efficiency transformers, and by enlarging their market share in China. The specific project objectives included drafting new minimum energy performance standards for distribution transformers and developing a total-cost-owning (TOC) tool based on life cycle costing methodologies to enable users of transformers to make transparent purchase decisions with energy-saving considerations. Furthermore, the project aimed to develop an eco-design standard for power and distribution transformers to influence all parts of the supply chain. The active engagement of many stakeholders was necessary to achieve this goal. This involved policy-makers concerned with energy consumption issues of the National Development and Reform Commission (NDRC) and the Standardization Administration of China (SAC), as well as the committees setting standards in the China National Institute of Standardization (CNIS), the China Electricity Council (CEC), and the China Electrical Equipment Industry Association (CEEIA).

It also included end-users such as power transmission, distribution utilities, energy-intensive industries and manufacturers of transformers, as well as mainly small and medium-sized enterprises, and energy conservation and supervision centres (ECSCs) in key energy-using provinces. This was not the first international cooperation initiative on transformers. In 2009, the World Bank and State Grid China Corporation signed a grid CDM project cooperation letter of intent with the intention to replace 166,000 high energy-consumption transformers. The SWITCH-Asia project took a wider perspective and sought to increase the market penetration of higher efficiency transformers and used various strategies to reach this goal. Developing energy efficiency standards for transformers was one major part. International and domestic project experts worked on standards for minimum energy performance, on eco-design and on a total-cost-owning (TCO) guideline. TCOs enable users to assess the total cost of an investment across the product life cycle, including the operation costs. With this set of mandatory and voluntary standards and tools the project tried to build more favourable market conditions for higher efficiency transformers. A second part of the project activities consisted of raising awareness among end-users of transformers, in particular power distribution utilities and energy-intensive industries. Such end-users are the leading energy consumers in China and pose a major threat to China's energy security. They urgently need to reduce their growth in electricity use. Energy savings in the use-phase of transformers are often not considered in the investment decision simply because more efficient transformers are more expensive. The project therefore helped energy managers and procurement officers of utilities to build up their capacity for undertaking life-cycle cost analyses. The TCO tools supported energy managers in procuring equipment. Furthermore, energy conservation and supervision centres in local government offices were in a strategic position to participate in the project efforts to create awareness and improve the compliance of energy-saving standards and policies. In total, the project held 15 seminars with about 3,000 participants from power utilities and industrial end-users engaged in training courses on TCO and energy policies. A third part of the project was capacity building with local SMEs in various provinces (including Henan, Hebei, Shandong, Guangdong and Jiangsu), to enable them to design and manufacture higher efficiency transformers. Only when manufacturers have this capacity, will they progressively expand their product portfolio of transformers and contribute to the transformation of the market. The project furthermore contributed to the implementation of China's macro-level energy and energy conservation policy framework. Efficient transformers as part of an optimised energy system contribute to the Medium and Long-term Energy Conservation Plan as laid out in China's 12th Five-Year Plan. The project addresses policy-makers including those at the Standardization Administration of China (SAC) and the National Development and Reform Council (NDRC), who are responsible for undertaking management, supervision and coordination of energy saving through more efficient equipment and technologies. The China National Institute of Standardization (CNIS), in charge of setting standards and reporting to SAC and NDRC, leads the work on minimum energy performance standards (MEPS) in the project, supporting the introduction of new national standards for efficient transformers. Through this action, the program will help to increase the mandatory minimum energy efficiency requirements in the long term, and effectively push inefficient transformers out of the market. The programme also contributed to the redesign of a mandatory energy efficiency label and a higher efficiency certificate for top performing transformers. For instance, the eco-design standard and TOC guidelines were all issued, becoming officially effective in 2012. These standards increase the minimum efficiency of transformers, and as this is a mandatory target, it imposes compliance for all newly installed transformers. Also in 2012, the State Council of China (2012) issued the Energy Conservation And Emissions Reduction's Twelfth Five-Year Plan, which outlines specific targets to reduce the load loss of transformers by 17 to 19 per cent. Furthermore, with a view to realising these targets, support instruments have been issued, such as a fiscal subsidy entitled 'Energy-saving products & people-benefit project implementing rules for the promotion and application of Higher Energy Efficient Transformers' in November 2012. Finally, these initiatives on standards will pave the way for efficient transformers to be included in the China Catalogue of High-Efficiency Products, normally used in public procurement processes of the local administrations. Products included in the catalogue can bring tax advantages for the production companies.

Finally, the project contributed to a transformation of the market by providing specific instruments for different target groups. It addressed manufacturers with a voluntary eco-design standard. The project partner, the China Electrical Equipment Industrial Association (CEEIA), was closely involved in ensuring that Chinese manufacturers accept the standard and subsequently highlight the benefits for the total supply chain. Efficient transformers will indeed only penetrate the market when end-users can assess the earnings from more efficient products. To enable end-users to take an informed decision, the project developed a total cost owning tool (TCO) and product database. With those tools, end-users can consider economic and technical data of the whole life cycle of a transformer prior to purchasing equipment. Overall, the project achieved a replacement level of 20 per cent of power transformers with the lowest efficiency by transformers of medium-level efficiency. This resulted in direct annual savings of around 887 million kWh, equalling emission reductions of about 846,300 tons CO_2 and 47 200 tons of sulphur dioxide in the five targeted provinces.

Summary, conclusion and recommendations

China's energy consumption trends have been on an upward trend fin recent decades and will likely continue to increase, posing a huge challenge to

China's energy security. Since China experienced steep rise in energy consumption in the early 2000s, energy efficiency is a major strategy for the country to ensure that it can meet the energy demands of industry, commerce and households. In particular, energy efficiency is the main strategy to protect industry from energy shortages and blackouts which impact on their productivity as well as on local employment. The policy developments in recent years show that China's energy-efficiency policies are becoming more sophisticated. Market-based mechanisms and other instruments are being applied more frequently, in addition to the traditional top-down approaches. The examples of the 1,000 Enterprises and 10,000 Enterprises programmes show that the Chinese government is adopting international instruments for industrial energy efficiency, tailored to the Chinese context, in this case voluntary agreement setting with industry stakeholders. This is a away of complementing the traditional top-down target setting approach. The 10000 Enterprises Programme also shows that improving China's energy efficiency will require targeting SMEs across all sectors, not only large state-owned enterprises. The two case studies of two key technologies designed to achieve higher levels of energy efficiency in industrial production, electric motors systems and transformers, demonstrate the complexity of achieving the uptake of more efficient technologies by industry. International cooperation for energy efficiency is an important means of addressing energy security, particularly when it comes to the exchange of policy instruments and harmonisation of industrial equipment standards. The experiences of previous international programmes promoting voluntary agreements and the SWITCH-Asia Programme in China show that transfer of both policy approaches and standard setting processes require adjustments to local conditions. Simple information or communication activities alone are generally not sufficient to achieve concrete improvements in SMEs. In order for them to have an impact, they must be supported by other activities such as capacity building, technical training and the provision of energy auditing services. Furthermore, the project experiences demonstrate the need for the ongoing structured engagement of industry stakeholders if further progress is to be made in energy efficiency. Finally, air pollution and health concerns of Chinese citizens who are suffering from extreme levels of PM2.5 pollution, are adding another dimension to the efforts required for industrial energy efficiency. The example of industrial energy efficiency demonstrates an increasing convergence of multiple issues including energy security, environmental security, people's health concerns and climate change. Governments and industries which neglect or even outright disregard these issues in their medium to long term decision-making are likely to face multiple risks to their local economies and to business in the coming decades.

Notes

1   Both case studies presented here are based on SWITCH-Asia Programme materials. Available at: www.switch-asia.eu/.

2 Detailed project information available at: www.switch-asia.eu/projects/china-mot or-challenge/.
3 The China National Institute of Standardization (CNIS) is one of the country's main bodies for providing technical expertise on energy-efficiency standards for national policy-making, therefore an important partner for international cooperation programmes. Available at: www.cnis.gov.cn.
4 Detailed project information available at: www.switch-asia.eu/projects/higher-effi ciency-of-transformers/.

Bibliography

Andrews-Speed, P. (2009) 'China's Ongoing Energy Efficiency Drive: Origins, Progress and Prospects.' *Energy Policy* 37: 1331–1344.

Cai, S. (2013) 'The Promotion and Application of Higher Energy Efficiency Transformers.' *China Electric Power Research Institute.* Available at: www.egeec.apec.org/dm sdocument/453 (accessed 14 January 2015).

Energy Conservation Law of the People's Republic of China (2007 Revision) Chinese and English version available at: www.lawinfochina.com/display.aspx?lib=law&id= 6467 (accessed 14 January 2015).

European Environmental Agency (EEA) (1997) *Environmental Agreements, Environmental Effectiveness.* Copenhagen: EEA.

Eichhorst, U. and Bongardt, D. (2009) 'Towards Cooperative Policy Approaches in China-Drivers for Voluntary Agreements on Industrial Energy Efficiency in Nanjing.' *Energy Policy* 37: 1855–1865.

Finnamore, B. and Davidson, M. (2011) 'China's Twelfth Five-Year Plan: Energy and Environment Positioned Strategically.' *China US Focus*, 29 March. Available at: www.chinausfocus.com/energy-environment/china%E2%80%99s-twelfth-five-year-pl anenergy-and-environment-positioned-strategically/ (accessed 1 February 2016).

Fisher-Vanden, K., Mansur, E. and Wang, Q. (2010) 'Costly Blackouts? Measuring Productivity and Environmental Effects of Electricity Shortages.' Berkeley, CA: University of California Institutions and Governance Program. Available at: http:// igov.berkeley.edu/content/costly-blackouts-measuring-productivity-and-environment al-effects-electricity-shortages%E2%88%97 (accessed 14 January 2015).

Fleiter, T., Eichhammer, W. and Schleich, J. (2011) 'Energy Efficiency in Electric Motor Systems: Technical Potentials and Policy Approaches for Developing Countries.' *Working Paper 11.* United Nations Industrial Development Organisation & Fraunhofer Institute for Systems and Innovation Research Karlsruhe.

GNESD (Global Network on Energy for Sustainable Development) (2010) *Achieving Energy Security in Developing Countries.* Risoe, Denmark: Global Network on Energy for Sustainable Development & UNEP.

Levine, M. D., Zhou, N. and Price, L. (2009) *The Greening of the Middle Kingdom: The Story of Energy Efficiency in China.* Berkeley, CA: Ernest Orlando Lawrence Berkeley National Laboratory.

Lian, R. and Serapio, M. (2013) 'China May Close More Steel Mills in 2014 to Tackle Pollution-Industry', *Reuters*, 16 December. Available at: www.reuters.com/article/2013/ 12/17/china-steel-output-idUSL3N0JV1NU20131217 (accessed 14 January 2015).

Lundqvist, D. and Mattsson, M. (2011) 'Effective Communication With SMEs on Energy Efficiency.' *Executive Summary. WGR 6.2 Core theme 6 Working Group Report 2.* Swedish Energy Agency & Motiva, Finland.

Moomaw, W., Yamba, F. Kamimoto, M. Maurice, L. Nyboer, J. *et al.* (eds), *IPCC Special Report on Renewable Energy Sources and Climate Change Mitigation*. Cambridge: Cambridge University Press. Available at: http://srren.ipcc-wg3.de/rep ort/IPCC_SRREN_Ch01.pdf (accessed 14 January 2015).

National Development and Reform Commission (2004) *Jie neng zhong chang zhuan xiang gui hua* [*Medium and Long-Term Energy Conservation Plan*]. Beijing: National Development and Reform Commission. Available at: www.china.com.cn/ chinese/PI-c/713341.htm (accessed 14 January 2015).

National Government of the People's Republic of China (2007) *Zhong guo ren min gong he guo jie yue neng yuan fa* [*China's Energy Conservation Law*]. Beijing: National Government of the People's Republic of China. Available at: www.china. com.cn/policy/txt/2007-10/29/content_9139273.htm (accessed 14 January 2015).

Price, L., Wang, X. and Yun, J. (2010) *China's Top-1000 Energy-Consuming Enterprises Program: Reducing Energy Consumption of the 1000 Largest Industrial Enterprises in China*. Berkeley, CA: Ernest Orlando Lawrence Berkeley National Laboratory.

Qiu, J. (2011) 'China Announces Energy-Saving Plans', *Nature News*, 4 March. Available at: www.nature.com/news/2011/110304/full/news.2011.137.html (accessed 14 January 2015).

Reinaud, J. and Goldberg, A. (2012) *Insights into Industrial Energy Efficiency Policy Packages: Sharing Best Practices From Six Countries*. Washington, DC: Institute for Industrial Productivity.

Reuters (2013) 'China's November Power Consumption Up 8.5 pct on Year', *Reuters*, 16 December. Available at: http://uk.reuters.com/article/2013/12/16/china -power-consumption-idUKL3N0JV1KY20131216 (accessed 6 January 2015).

Reuters (2014) 'Non-Fossil Fuels Rise in China's Energy Mix – Paper', *Reuters Beijing*, 14 January. Available at: http://uk.reuters.com/article/2014/01/14/china -energy-idUKL3N0KO1FA20140114 (accessed 6 January 2015).

Song, R. (2013) 'China's New Energy Consumption Control Target.' *ChinaFAQs* (15 March). Available at: www.chinafaqs.org/blog-posts/chinas-new-energy-consump tion-control-target (accessed 14 January 2015).

State Council of China (2012) *Guo wu yuan guan yu yin fa jie neng jian pai 'shi er wu' gui hua de tong zhi* [*Energy Conservation And Emissions Reduction's Twelfth Five-Year Plan*]. Beijing: State Council of China. Available at: www.gov.cn/zwgk/ 2012-08/21/content_2207867.htm (accessed 13 January 2015).

SWITCH-Asia Network Facility (2012) 'China Higher Efficiency Power and Distribution Transformers Promotion.' *Project Impact Sheet. SWITCH-Asia Programme*. Available at: www.switch-asia.eu/projects/higher-efficiency-of-transform ers/ (accessed 1 February 2016).

US Energy Information Agency (EIA) (2014) *China Country Data*. Available at: www. eia.gov/countries/country-data.cfm?fips=CH#elec (accessed 14 January 2015).

Wang, Z., Zeng, H., Wei, Y. and Zhang, Y. (2012) 'Regional Total Factor Energy Efficiency: An Empirical Analysis of Industrial Sector in China.' *Applied Energy* 97: 115–123.

Wu, F., Fan, L. W., Zhou, P. and Zhou, D. Q. (2012) 'Industrial Energy Efficiency With CO2 Emissions in China: A Nonparametric Analysis.' *Energy Policy* 49: 164–172.

Xinhua (2013) 'China's Cabinet Approves Energy Consumption Control Target', *Xinhua News Agency*, 30 January. Available at: http://news.xinhuanet.com/english/ china/2013-01/30/c_132139588.htm (accessed 6 January 2015).

Xue, Y. and Bressers, H. (2010) "Potentiality-Capability' Analysis Framework of Negotiated Agreements: A Comparative Study of Industrial Energy saving in the Netherlands and China.' Paper presented at the 16th International Sustainable Development Research Conference 30 May – 1 June, Hong Kong. Twente Centre for Studies in Technology and Sustainable Development, University of Twente, The Netherlands.

Yang, M. (2007) 'Raising China's Electric Motor Efficiency'. *European Council for an Energy Efficient Economy (ECEEE) 2007 SUMMER STUDY • SAVING ENERGY – JUST DO IT!* Available at: www.eceee.org/library/conference_proceed ings/eceee_Summer_Studies/2007/Panel_7/7.014 (accessed 14 January 2015).

Yu, S., Zhou, L. and Li, C. (2013) 'China Wrestles With Power Shortages', *Powermag*, 5 January. Available at: www.powermag.com/china-wrestles-with-power-shortages/?printmode=1 (accessed 14 January 2015).

Zhang, J. (2011) 'China's Energy Security: Prospects, Challenges and Opportunities.' *The Brookings Institution*, July. Available at: www.brookings.edu/~/media/re search/files/papers/2011/7/china%20energy%20zhang/07_china_energy_zhang_paper (accessed 14 January 2015).

Zheng, J. (2013) 'Hebei to Limit Air Pollutants, Shut Down Obsolete Plants', *China Daily*, 28 November. Available at: www.chinadaily.com.cn/china/2013-11/28/content_17136370.htm (accessed 14 January 2015).

11 The energy transition and energy security of cities

The urban dimension of Chinese energy issues

Giulia C. Romano

Introduction

The 'classical' vision of energy security often neglects the role that cities play as primary actors facing the challenge of ensuring adequate levels of energy supply as well as rationalising existing resources. In order to 'fill this blank', recent works on energy security have considered energy efficiency, decentralised energy and the topic of sustainable development as important components of the definition of energy security concepts, going beyond the traditional focus on the international geopolitical and commercial dimensions of energy supply, as well as on the availability of national supplies and the production strategies of governments (see Gheorghe and Muresan 2011; Sovacool 2011; Yergin 2011). However, a look at the content tables of several works that are specifically dedicated to energy security – either on international energy security or focusing on the situation of China – shows that very few works have made the link between energy security and the contribution of cities explicit.[1]

The concepts of 'sustainable urban development' and, more recently, of 'low-carbon development' inherently include considerations on both energy consumption and energy supply, namely by stressing the great importance of enhancing energy saving performances while replacing fossil energy sources with renewable energy. The current nexus between the dimension of 'sustainability' and the need to respond to challenges brought by energy security and emissions of CO_2 and other pollutants is already evidenced by several works (see Fitzgerald 2010; Rydin 2010). However, in order to contribute to an extension of our vision on energy security we felt we should clarify the link between national and local roles. We are hence adding an urban focus to the literature on energy security.

Although the connection may seem obvious, broadening the concept of 'energy security' in order to encompass the contribution of policies and energy choices at the city level appears to be inevitable, especially in a context like China, where energy shortages and environmental pressures have become increasingly important problems. Moreover, the growing local demand, driven by both urbanisation and industrialisation, requires us to turn our attention to consumption patterns at the city level. We will thus begin by discussing the importance of this linkage between cities and energy security. Through an analysis of current consumption patterns and results of existing policies towards limiting energy consumption in urban areas (in particular in the building and transportation sectors), we will highlight an important inconsistency between national policies and local choices with regard to energy use, one that constitutes a potential threat to China overall energy security strategy. Finally, we will conclude by discussing energy security and sustainable urban development in light of the current environmental and energy crises.

The energy transition of cities as an energy security strategy

When discussing energy security and national policies that ensure a stable supply of energy sources, one cannot help but consider the role that cities play in complementing these policies and mobilising concepts such as 'energy transition'. Coined by researchers of the Öko-Institut in 1980, the *Energie-Wende* ('energy transition') proposes a progressive abandonment of traditional energy sources, namely fossil fuels, as well as of nuclear energy. In particular, it implies a combination of renewable energy, energy efficiency and sustainable development, together with the progressive decentralisation of energy production.[2] In this vision, energy policies should consider reorienting their attention from demand to supply, shifting from a centralised provision of energy to a decentralised offer and tackling energy wastes thanks to energy efficiency and energy saving measures. By helping localities to reduce consumption and aggregate easing countries to decrease their reliance on external energy sources, such a switch would therefore contribute to energy security.

The urban dimension of energy security demands special attention, and concepts such as the 'energy transition' of cities are of particular importance, especially in a context of rapid urban growth like China, where energy consumption of cities has already reached 75 per cent of the country's total consumption and is expected to reach 83 per cent by 2030 (Kamal-Chaoui and Cointreau 2014: 30). A focus such as this should look not only at growing energy demand and potential shortages at the core of energy security concerns, but also at the intrinsic environmental dimension linked to the consumption of fossil fuels. This consumption is expected to produce 85 per cent of China's energy-related greenhouse gas emissions (World Bank 2010: 31) and weight for 20 per cent of the world's total energy consumption and one quarter of the world's total oil demand (McKinsey Global Institute 2009: 18). In order to feed this increasing expansion of China's urban areas, the country will need to grow from 700 to 900 GW of new coal-fired power capacity (ibid.).

Seen from a carbon emissions perspective, cities at prefectural levels and above[3] already accounted for 55 per cent of China's total greenhouse gases emissions in 2009, with the first 100 ranking cities covering 41 per cent of the emissions (Niu 2009). In 2008, when the population of Chinese cities did not reach half of the country's total population – at 46 per cent – its greenhouse gases emissions had nevertheless already reached 90 per cent of the country's total (Guo and Liu 2010: 22). A study made by a Chinese research team on the connection between the growth of urbanisation and the increase in emissions shows that since the opening up of the country in 2008, there is a strong correlation between the two progressions (correlation coefficient: 0.9731), meaning that urbanisation can definitely be seen as one of the main factors of China's increasing carbon emissions (He 2010:5).

With these projections in mind, either in terms of energy or the environment, Chinese cities, and especially their physical expansion, certainly matter. The Chinese government plans to raise urbanisation to 70 per cent by 2035,[4] implying the move of around 350 million people from rural areas to cities (Démurger 2010), as well as an increase in the demand for steel and cement to construct new dwellings and to provide more infrastructures. In 2009, Chinese steel consumption reached 46.4 per cent of the world's total steel consumption, while cement reached 48.7 per cent (Guo and Liu 2010: 24). An increasing need for the large-scale construction of facilities, new buildings and new residential areas will put further pressure on the current energy demand, as urbanisation in China is currently not only the main producer of GDP, but also the main 'energy-eater' (*ibid.*). As a World Bank report on the role of cities in the climate change issue highlights, China is the sole country expected to obtain 90 per cent of its GDP 'from urban areas in 2025, but many of the associated buildings and large-scale infrastructure have not yet been built' (World Bank 2010: 31). Within the next ten years, the GDP produced by urbanisation will rise from the current 75 per cent to 95 per cent (McKinsey Global Institute 2009: 15).

The energy shortages and environmental degradation of cities in the energy security equation

Urbanisation and an increase in the energy-greedy industrial sectors such as steel and cement production, coupled with a reduction of controls on energy consumption, led the country to witness important episodes of power shortages, starting in 2003 (Andrews-Speed 2012: 20). The emergence of these episodes can be attributed to a series of interlinked factors, foremost amongst which were artificially low electricity prices (and energy tariffs in general), insufficient capital investment in the power sector, limited capacity to enhance generating equipment availability (Lam 2005), and the easing of controls over energy consumption. These episodes, which had an impact on industrial production, were very serious and imposed on the country a necessary reconsideration of its energy management. In the second half of the 2000s, the Chinese government indeed placed energy efficiency and energy conservation as a national priority by issuing a new strategy – the Medium and Long Term Energy Conservation Plan – establishing the target of reducing energy intensity by 20 per cent within 2010 (Andrews-Speed 2012: 20). Energy-saving objectives were also included in the 11th Five-Year Plan (2006–2010) and again in the 12th Five-Year Plan (2011–2015), showing how the central government policies had prioritised the subject.

To highlight the weight of energy conservation in the government's energy strategies, the reduction objectives were declined at the provincial level and were coupled with a strict control of their achievements. Beginning with the 11th Five-Year Plan, the evaluation of performances of local leaders is indeed now also based on energy conservation abilities (Qi 2013: 234). However, far from being solved just by an increased attention to energy conservation, the problem of energy shortage in cities remains a great concern. For instance in the autumn of 2013, the government had to make a choice between supplying industries or supplying households (Huang 2013) when it was faced with an important gap between the demand and supply of gas. Of course the decision that was made and that continues to be made when the situation arises, is to sacrifice industrial needs for the benefit of residential users, a choice that is also dictated by the pre-existing strategy of the Chinese government to restructure the industrial sector, as many sectors of production are being characterised by redundancies and overproduction. As gas also became an important element in the depolluting strategy of Chinese cities, for example by replacing coal in electricity production and being encouraged to use it as an alternative fuel for vehicles, the urge for cities to cope with shortages and to reduce energy waste became of primordial importance for China's energy security concerns as a whole.

The growing energy demand, pushed by the expansion of Chinese cities both in spatial dimensions and in population, started to become an issue for energy security in terms of its environmental aspects as well, particularly evidenced by air pollution crises. The deterioration of the quality of air in Chinese cities dates back to the heavy industrialisation period of the Maoist regime, rendering places such as Shenyang, meaning 'rust-belt' in ancient Chinese, one of the unhealthiest places in the country (Larson 2011). In recent years, air pollution has taken emergency characteristics, regularly appearing in Chinese cities as soon as the cold starts to hit. The image of the thick haze covering Beijing and other northeastern Chinese cities became a symbol of such environmental crises. Despite the establishment of air protection policies with the first Chinese environmental law (1989), air quality in Chinese cities has continuously registered significant deterioration over the last 20 years. In 2001, a World Bank report listed 16 Chinese cities among the world's 20 most polluted (World Bank 2001: 174, cited in Economy 2005). Beijing had already been included on the list. In 2002, a survey conducted by the then State Environmental Protection Agency (SEPA, now Ministry of Environmental Protection) on the state of air quality in over three hundred Chinese cities indicated that the total concentration of suspended particulates of almost two thirds of the cities were well beyond the minimum level of acceptability established by the World Health Organisation (Economy 2005: 72).

As for the capital city, Beijing adopted certain measures to improve the air quality for the 2008 Olympic Games. The city government decided to relocate polluting factories beyond the city borders and decreed a temporal suspension of industrial production in surrounding provinces. Polluting vehicles were also banned, while the circulation of private cars was restricted. However, the possibility of seeing blue skies in Beijing was not granted for that long. Contrary to expectations of an improvement in the quality of the air surrounding the capital, the trend took a negative turn in the years following the Olympic Games. In winter 2011, the American Embassy's diffusion of data on PM 2.5 concentration in Beijing's air again brought the problem of heavy polluted fogs to international attention. The serious crisis registered in January 2013, in which PM 2.5 concentration reached the peak of 886 micrograms per cubic metre,[5] deserved the appellation of 'air-pocalypse' (Romano 2013: 88). Beyond Beijing, the thick haze also covered other surrounding areas (Tianjin, Harbin and the northeastern provinces of Shandong and Hebei to name but a few), and the phenomenon then became more important in other localities that were not normally very interested in heavy air pollution, like Shanghai and Jiangsu provinces.[6] The issue surely adds doubts regarding the current depolluting and air control policies, and further urges the need to find cleaner solutions to coal usage, although this resource is still going to be dominant in the years to come. Moreover, the use of coal for its transformation into gas or other energy sources is going to create a heavy burden to the already weak water resources of the country, further highlighting the important link existing between energy security and the environment (see Zhi and Wang, this volume, Chapter 2). The perspective of reducing energy consumption for the sake 'of their own lungs' puts a big pressure on Chinese cities to adopt measures of energy sobriety by limiting excessive increases of their demand.

As has been presented in Chapter 9, the adoption of decentralised supply strategies through the implementation of decentralised energy management still currently remains a difficult task for China, where the power of great central energy producers fundamentally discourages the development of a decentralised energy supply. The realisation of energy transition inherently includes a significant restructuration of the decision-making system on energy choices, shifting production and management from the hands of few centralised and oligopolistic players to the responsibility of cities or smaller agglomerations. This transformation of local communities in important stakeholders of their own energy provision would entail a progressive 'democratisation' of energy supplies eschewed by powerful energy groups. Moreover, the artificially low electricity prices fixed by the government – especially coal-fuelled electricity – make substituting cleaner resources (renewable energy) an undesired option. Finally, despite the presence of some experiments introducing renewable energy supply at the city level[7] in China, as well as in other countries, cities have limited competence in decision-making in the energy field, traditionally a prerogative of the central government. With regard to energy management, the room for manoeuvre left to Chinese cities is limited to intervening in certain sectorial areas such as transportation, construction, space development and dealing with industries under their administrative competence. Chapter 10 addresses the industrial sector, whilst this chapter chooses to focus its attention on issues and policies that concern the construction, urban planning and transportation sectors. The increasing area that buildings (and their location) and transportation (and roads construction) cover in China's current and future energy demand make them entirely eligible to enter into a comprehensive energy security concept.

Opportunities and facts regarding 'energy transition' in Chinese cities

Reducing the 'energy hunger' of old and new buildings

The impressive spatial development of Chinese cities and newly urbanised areas can clearly be seen when travelling through the country, and especially its cities. Hosting numerous construction sites, big cities like Beijing, Shanghai, Shenzhen and Chengdu, with their increasing suburban agglomerations, as well as other medium or smaller cities in China, can be said to be constantly 'changing their face'. Old buildings are demolished, only to be replaced by new residential or commercial areas that are often in the form of high-rise constructions, significantly modifying a city's landscape within the space of just a few months. New roads join the already considerable provision of ring roads and highways, bringing cities like Beijing to further expand on road infrastructure, up to a seventh ring road even (Chinadaily 2014). It is practically impossible to miss the outline of cranes in China's landscapes or to notice the presence of the numerous construction sites where residential towers 'grow' with remarkable rapidity. Far from the borders of cities where roads are not yet built, it is possible to observe construction sites in the middle of nowhere, already a tell-tale sign that in a short space of time they will be entirely covered in new roads and buildings. In these dramatically fast-growing spatial expansions, rural land is progressively seized, while the demand in energy increases.

Urban expansion passed from 0.81 billion square metres in 2000 to 2.77 billion in 2010 (in terms of new floor space per year), increasing 3.44 times in just 10 years (Shui and Li 2012: 34). In terms of total floor space, urban expansion reached 33.3 billion square metres in 2011 (World Bank 2014: 501). Estimates in 2012 foresaw an extension of new floor space from ten to 15 billion square metres in residential buildings, and ten billion square metres in public buildings by 2020 (Shui and Li 2012: 34). Against this background and beyond the important considerations in energy and environmental terms that such colossal construction would imply, the energy conservation and energy efficiency of buildings is of paramount importance.

The US Energy Information Administration (EIA)'s 2013 report entitled 'International Energy Outlook' shows that in 2010, the energy consumption of China's residential sector covered approximately 10 per cent of the country's final energy consumption (EIA 2013a). According to EIA forecasts, energy use in this sector will grow by 3.6 per cent per year by 2040, becoming 'almost twice as high as in the United States' and reaching 24 per cent of the world's total residential energy use. As the report highlights, such an important increase of the energy consumption in this sector is surely linked to the speed of urbanisation and the following 'annual additions of new buildings', coupled by fast economic growth and the general improvements in living standards. However, given the lack of enforcement of building regulations, the speed of construction, and often the use of low quality materials, the overall quality of many Chinese buildings is low, giving them quite a short life span (20 to 30 years) compared to buildings in Europe or in the US (70–75 years) (Larson 2012). At the same time, notwithstanding the presence of regulations, energy-saving measures like simple building insulation are often lacking or at least poorly applied (den Hartog 2010: 66).[8]

Energy efficiency in building codes and standards

Building codes are quite particular in China following the virtual division of the country into different climatic regions and through the 'Qinling-Huai' line. Virtually traced across the country from the Western group of mountains of Qinling to the Eastern River Huai, this imaginary line was adopted in 1949 as a demarcation to provide heating to buildings. This means that areas below the 33° parallel are not entitled to heating systems. However, these 'Southern' areas do not all equally present the 'warm winters, hot summers' characteristic of provinces like Guangdong and Guangxi. Southern areas were thus further divided into areas of 'cold winter, hot summers', including the city of Shanghai, and areas of 'warm winters, hot summers'. Building energy efficiency codes is relevant for three of the five Chinese climatic zones, namely those of 'cold winters, warm summers', 'cold winters, hot summers' and 'warm winters, hot summers'.

In 1986 and 1995, the Chinese government indeed promulgated energy efficiency standards for buildings in the Northern area, mandating a 50 per cent reduction of energy consumption compared to buildings constructed in the 1980s, when no energy efficiency measures in the envelope and equipment were applied. In 2001 and 2003, new codes for the non-heated areas ('cold winters, hot summers' and 'warm winters, hot summers') were issued in order to enhance the energy efficiency of buildings in these zones where air conditioning prevails as the main heating system. Moreover, following the international trend of creating specific standards for 'green buildings',[9] China also adopted its own rating system, the so-called 'three-stars', in 2006. This system, in many ways similar to the American LEED system (Leadership in Energy and Environmental Design) for sharing similar rating categories,[10] further adds an 'operation and management' dimension that is deemed to be very important in Chinese constructions. Indeed, since compliance remains a significant challenge in China, buildings are evaluated one year after the operation, whereas the US LEED system usually certifies the building by evaluating it after the design phase. Nevertheless, despite this equipment of regulatory tools to increase the energy saving potential of buildings, codes still lack enforcement and implementation in China.

Since the heat billing system remains substantially untouched, the introduction of codes for buildings in the Northern Areas did not end up translating into a thorough application, whereas in the South, low electricity prices make the extra cost of the addition of energy efficiency measures in buildings to be reimbursed in 50–100 years. Neither building constructors nor building owners and users are incentivised to implement measures towards reducing the energy consumption in buildings. Beyond the lack of enforcement of building codes, the low efficiency of heating systems and the inadequate insulation of building envelopes (the exterior walls) make energy consumption for heating still 2 to 4 times higher compared to Northern European houses that present similar climates (Cai *et al.* 2009: 2055).

In China, several factors must be considered when observing the poor results in energy efficiency of buildings and the energy savings that can be obtained. First of all, the energy saving target foreseen by the existing codes are yet to be sufficiently ambitious. In 2010, the codes were revised when the energy saving target was raised from 50 per cent to 65 per cent (Shui and Li 2012: 22). However, since the baseline taken by these codes is the standard that was set in the 1986 code, where energy saving measures were almost absent, the energy-saving performance of these buildings still remains significantly low compared with the minimum standards required in European countries.[11] Secondly, the current monitoring system is still too weak to ensure any enforcement of these codes. Although the Chinese Ministry of Housing and Rural Development (MoHURD) affirms that the compliance with codes and standards has improved in recent years through more rigid inspection controls, increasing from 53 per cent and 21 per cent (design stage and construction stage respectively) in 2005 to 99.5 per cent and 95.4 per cent in 2010 (ibid.: 40), the conditions for a wider implementation of new energy efficiency standards are not yet met. Significant limitations still exist in the supervision of construction sites, as the local governmental agencies that were appointed for inspection lack skilled staff to carry on widespread controls. Sometimes inspection itself even lacks and 'green-washing' becomes quite common practice.[12] The codes themselves are not detailed enough to provide guidance on measures towards achieving energy efficiency in buildings through the design phase, while the rapidity of the design process does not give professionals enough time to include the parameters technically enabling them to take energy efficiency concerns into account.[13] Moreover, during the various phases of the design process, it is not possible to guarantee that the measures proposed by architects are followed during the stages of the building approval process.[14] Finally, insufficient control over both the construction process and unskilled workers (often recruited among migrant workers) can explain how energy-saving measures, and other designed measures, are lost along the path of building construction.

Beyond the difficulties that exist in the 'building production chain', in which it is possible to include the difficulties of certifying whether the building materials chosen really guarantee energy saving effects, the cost of enforcing codes is still deemed to be too high if energy prices are taken into account. As previously mentioned, the payback periods for the investments carried out in energy saving – through improved building envelopes, for instance, or through using energy-saving electric appliances – are still too long for potential buyers to consider energy-saving as an important element in the definition of their preferences (Richerzhagen *et al.* 2008: 65). Moreover, one must take the building construction market in China, which is mostly occupied by big real estate developers, into account. This situation creates a split incentive problem, since developers are more interested in saving money in building construction while obtaining the highest profit from their sales than in investing a large amount of money in energy-saving materials or in meeting the requirements of design. In the case of the green buildings rating system, notwithstanding the existence of subsidies accompanying the achievement of the two-star and three-star levels, the costs that developers must bear, as well as the uncertainties linked to the process of being granted the subsidies, instead make them refrain from introducing those measures.[15] Finally, there is little the consumer can do in terms of building energy efficiency if not by choosing the type of electric appliances. However their high prices, compared to electricity costs, thus the payback period, continue to discourage buyers.

With regard to building operations, applying measures to reduce the energy intensity of building operations in China is capable of saving up to 29 per cent of energy 'at virtually no cost' (ibid.: 26). Indeed, most of the energy used for space heating of urban residential and commercial buildings is wasted, especially in buildings that were constructed before the code adoption. A report issued by the World Bank in 2001 showed that residential buildings in China could consume between 50 and 100 per cent more energy for space heating than buildings in Europe or the USA that were under the same climatic conditions. Beyond being less comfortable than buildings in the West, their heat loss through exterior walls – 'the greatest single source of heat loss in these buildings' – was from three to five times higher than similar buildings in the Northern regions of Western countries (World Bank 2001: 1). Moreover, as far as heating is concerned, buildings were constructed with high-energy inefficient design principles. For instance in northern areas, where buildings are provided with central heating systems, their obsolete design, which still conforms to models developed in the 1950s, renders it impossible to adjust the indoor temperature by the single household, resulting in significant energy waste. For householders, the only way to regulate the temperature is to open the windows, constituting an unreasonable waste of energy that could be tackled through enhanced design and by adding the possibility for the single household to adjust the temperature according to its heating needs.

From the picture drawn so far we can observe that buildings in China present important problems in terms of energy efficiency, though the average energy consumption of Chinese inhabitants is still more parsimonious than that of European or American households, and the energy consumption of buildings remains, for the most part, lower than that of more comfortable Western dwellings. The United Nations database for the 2011 energy consumption per capita in kg of oil equivalent shows energy consumption values of 2,029.4 for China, 3,867.5 for France, 3,811.5 for Germany and 7,032.3 or the USA.[16] However, with increasing living standards and related requirements for more comfort, energy consumption levels in many Chinese households are becoming comparable to European or North American ones, and therefore surely require an enforcement of building codes, an enhancement of energy-saving standards for new buildings, a widespread diffusion of energy-saving appliances, as well as energy-saving behaviours.

As the World Bank report already pointed out more than 10 years ago (ibid.), the dimensions of new building constructions in China make this country's case rather unique. If the inefficiencies along the construction process are not soon addressed, the risks that China is going to face in terms of energy security are considerable. Indeed, if current energy waste levels are maintained through a weak application of existing codes through the obsolescence of their energy-saving targets and the much too artificial low energy prices, the extent of energy waste is likely to render energy shortages more frequent and more important given the preponderant place urbanisation will have in the energy consumption of China's near future.

'Thirsty' cars and sprawling cities

Apart from buildings, two other important factors that already impact – and will increasingly influence – energy security and environmental matters are the transportation sector and the current patterns of urban development. Since they are strictly connected, these two factors have been put together. The rapid expansion of cities is increasing the need to enhance the public transportation fleet, pushing municipalities to invest in road infrastructure and buses and rail transportation. According to McKinsey's report, already 51 per cent of investments in urban infrastructures was dedicated to road, bridges and mass transit between 2001 and 2005, the rest being invested in public utilities (McKinsey Global Institute 2009: 377). Following the pace of urban expansion, the need to provide more roads is also an imperative given the current diffusion of private-owned cars. Between 1990 and 2005, private-car ownership increased by an annual 31 per cent, while the construction of new roads only reached an increase of 10 per cent per year (ibid.: 378). Moreover, due to the demand stimulation promoted by the Chinese government during the 2008 and 2009 years of the global financial crisis, vehicle sales increased by 46 per cent in 2009 and 32 per cent in 2010 (EIA 2013b). This discrepancy between the increase of private-car diffusion rates and the space for road construction led to the already notorious phenomena of congestion. Driving through big cities such as Beijing and Shanghai has progressively become a complicated issue, where traffic jams are a given at almost every moment of the day. Particularly in Beijing, notwithstanding the extensions of ring roads following the expansion of the city and of the car fleet, traffic deterioration and congestion are perceived to be worsening.

It is not possible to provide an exact estimation of the total energy consumption of the transportation sector because it is only partially documented by official statistics. According to a study on energy consumption in the transportation sector, data was only available for the public sector, and not for the private (IFEU 2008: 6).[17] However, some elements could be found in different reports. For instance, according to the US Department of Energy, as far as oil consumption is concerned, 40 per cent of oil imported by China is used for transportation, of which two per cent goes to private cars (US Department of Energy 2006). The report points out that this percentage might reach 50 per cent by 2020, while oil needed for privately owned cars will reach 10 per cent. Moreover, the growth of the transportation sector is also seen as one of the major drives for China's greenhouse gas emissions, as well as of main air pollutants. If compared with the available data for 2011, CO_2 emissions from transport have almost sextupled since 1990, reaching 623 millions of metric tons per year (from 109 million in 1990).[18] Although current emissions from the transportation sector only account for 10 per cent compared to the average 25 per cent of OECD countries (Pan and Liu 2011: 225), the expected increase of the transportation sector after urbanisation and increasing wealth, which will not stop the purchase of private cars, will surely significantly raise this percentage.

An unsustainable increase in the number of private cars

The increasing number of privately owned cars brought China to become the largest automotive market in 2009 and, according to forecasts, energy demand for the transportation sector will grow at an annual rate of 3.7 per cent by 2040 (EIA 2013b). This growth in demand will be due not only to the expansion of the car fleet in China but also to a more generalised growth of the transportation sector. In terms of the demand of fuels, China has already become the major driving force of the world's growth in demand and the second largest consumer (behind the USA). China's share of global liquid consumption for transportation reached 9 per cent and is likely to grow up to 20 per cent by 2040. Oil could be replaced by gas, which has already been promoted by the Chinese government in its depolluting strategy. Indeed, at the end of October 2012, the Chinese government issued a new Natural Gas Utilisation Policy, beginning in December 2012 and clearly hoping to promote cleaner transportation and logistics through the use of liquefied natural gas (LNG). The then existing policy, dating back to 2007, divided gas consumers into four categories: 'preferred', 'permitted', 'restricted' and 'prohibited'. In the updated 2012 version, vehicle categories have been further expanded and detailed and now include buses, taxis, logistics distribution vehicles, passenger vehicles, sanitation vehicles, trucks and vessels (V&E Law 2013). Through this policy, the Chinese government expects the transportation sector to reach 65 per cent of total LNG use by 2015, while when the policy was issued it was just at 25 per cent (ibid.).

On a pure energy security discourse, under the current deficit in gas supply the strategy of replacing oil with gas does not provide an answer, although it would certainly respond to the pressing problem of air pollution in cities. However, the recent gas shortage in the winter of 2013 already significantly highlighted the difficulties of such a strategy, further urging the country to speed up the construction of its LNG terminals and to increase the supply through importing more gas from pipelines. Beyond the issue of dependence from foreign supplies though, there are also cost problems that do not make gas a viable option for the big energy companies of China unless the country's government significantly increases current sale prices. As seen in Chapter 2, important energy companies in China are losing money when they import gas from Central Asia, and LNG is bought at very high prices. Under a perspective combining both environmental concerns and energy security, a fitting strategy would imply the discouragement of the use and diffusion of privately owned cars while extending the offer of public transportation systems. However, notwithstanding an increasing provision of modern public transportation systems in Chinese cities – from efficient metro lines to BRT systems – the growing wealth of Chinese families is not likely to stop the trend towards a larger diffusion of cars, still considered to be a status symbol.

Since 2004, the government has put strong emphasis on public transportation. At the time, the Ministry of Construction (now MoHURD) issued guidelines for public transportation development, further endorsed by the State Council, which underlined the priority of urban public transportation (Pan 2011: 10). Three years later, public transportation was listed among the priority areas to promote energy conservation. Public transportation was mentioned for the first time in an official document issued by the State Council entitled the Comprehensive Energy Reduction Work Programme, linking the construction of public transportation services to the energy-saving strategies in urban transport (ibid.). Accompanying this prioritisation of public transportation in order to discourage the diffusion of private cars, some cities also issued important policies such as vehicle registration plate auctions in the city of Shanghai. However, if this move managed to strictly regulate the provision of car plates in Shanghai city on the one hand – resulting in very reduced numbers compared to Beijing[19] – it did not on the other hand really significantly control the total growth of private cars in the city. Other than the desire of Shanghai families to own a car, they found other ways to be satisfied, for instance by getting licence plates from nearby provinces. In Shanghai it is indeed possible to observe many cars with number plates from Jiangsu or Zhejiang provinces.

Furthermore, taking the perspective of an industrialising country, car production is not likely to be replaced, much to the benefit of other types of production. The car sector remains a symbol of a country's engineering and technical capacities, an important job provider for the whole industrial chain and finally a symbol of increased status. In other terms, the increase in the number of privately owned cars is unlikely to stop. Moreover, despite the presence of strong incentives to promote the use of electric vehicles and the progressive installation of terminals for battery recharging, the diffusion of this type of transportation means has certainly been limited. In 2010 for instance, the Chinese government issued a policy to promote the diffusion of electric vehicles, establishing an ambitious target of 500,000 new car sales per year, making this sector one of its new emerging strategic industries. According to its ambitious plans, China should reach five million electric cars by 2020, backed by subsidies directly addressing car manufacturers. This policy, which involves five main cities – Shanghai, Changchun, Shenzhen, Hangzhou and Hefei, also known for being China's main car manufacturers – offers 60,000 Yuan for an all-electric passenger vehicle and 3,000 Yuan for private buyers. However, the diffusion of electric vehicles by far remains limited. Customers complain about the high prices of battery-powered models, the insufficient supply of charging stations and the concerns for safety (Tian 2013). Nevertheless, whichever type of fuel is used – oil, gas or coal for electricity – the linkage between the increase in cars and energy security still remains an important problem, even if the country does switch to electric vehicles. The increase in the number of vehicles will not only need more energy, but will also require more space for roads and parking, putting further pressure on Chinese urban spaces.

Wide roads and highways vs. human-oriented spatial planning: patterns of 'modernisation' in Chinese cities[20]

An environmentally sound and energy-efficient urban strategy can also be achieved through a better design of urban form. There is a general agreement that compact cities are more sustainable than sprawling ones, since by increasing building density and reducing the width of roads, distances will also become reduced, thus reducing the need to commute (Naess 2001; Rydin 2010; Baeumler *et al.* 2012). Be that as it may, this argument is not completely shared (see for instance Breheny 1995). Abandoning functional zoning – distinguishing residential areas from working or commercial areas – and promoting mixed use would also play an important role in discouraging car use for daily commuting. Hence a careful consideration of urban design and patterns of mobility is a crucial element in an urban strategy that is willing to tackle excessive energy consumption and its related pollutant emissions.

In China, planning practices of cities and the expansion of urban areas followed different patterns according to different phases of China's contemporary history. During the era of centralised planning, cities were built according to the Soviet model of urban planning, basically serving the purpose of industrial development (Wu 1999). Under this pattern, urban planning was highly centralised, following a two-tier planning system that first defined the coordination of urban spaces with factories at a master plan level, and the coordination of factories and workers' villages at a level of detailed layout plans. It was principally serving economic planning needs, mainly playing a site-selection role for the placement of factories. Urban settlements needed to be transformed into 'production cities', which made their development linked to the construction of new industrial bases. The legacy of this type of urban organisation can be seen in the big squares and grand avenues that characterise cities such as Beijing, and which serve parade events, the presence of work-unit apartment blocks, and the low occupancy rate of space of these constructions within the compound (a moderate 20 per cent), resulting in important phenomena of urban sprawl (ibid.). The rupture of Sino–Soviet relations and the Cultural Revolution further exacerbated the problem of urban sprawl. The practice of city planning was first progressively abandoned and then completely dismissed. During those years, urban planning administrations were cancelled, urban planners were sent to the countryside for re-education and urban development was encouraged to take place in whatever vacant space could be found (ibid.: 177–178).

With the opening up and reform, the development of Chinese cities started to follow new patterns that resulted in more encouraged urban sprawl. These years were characterised by a progressive decentralisation of decision-making and fiscal powers from central to local governments, as well as by the marketisation of urban land-use rights. Changes in the economic production patterns characterised by a reduction of the importance of state-owned companies, together with the privatisation of urban enterprises, the 'tertiarisation of the urban economy' and the progressive increase of foreign investments in certain cities further impacted on the shape and face of Chinese cities (Ma 2004: 242). Two important reforms triggered post-reform Chinese urban development: the 1986 Land Administration Act, which made it easy for municipal governments to sell land-use rights, and the 1994 Tax Sharing System, which established a separate collection of certain types of taxes for the central and local governments. Introducing the possibility to sell land-use rights, the Land Administration Act made it possible for local governments to convert rural lands to urban use.[21]

The fiscal reform then introduced important challenges for city governments and obliged funds to be raised in order to finance the construction of infrastructures for the growing needs of cities. The consequence of such a reform is the increase of their debt, whereby municipalities indirectly borrow loans from commercial banks or the China Development Bank through municipal government-owned urban development investment corporations (the so-called UDICs[22]), and there is a progressive seizing of rural land, the only resource local governments can really dispose of. Through the possibility of converting rural land into urban land offered by land acquisition plans, city governments can obtain rural land at very low prices that are set by the state (including a calculation of agricultural revenues and of householders' relocation costs).[23] Once provided the necessary infrastructures, cities can then sell or auction the converted land to developers. The serviced land-use rights can be sold at very high prices, especially when land is used for commercial or residential purposes. At the end of 2013, the average price for residential land in major cities was 5,033 Yuan per square metre and 6,306 Yuan per square metre for commercial use (Fung 2014).

There are objective needs for cities in continuous expansion to seek for land in order to meet growing housing demands. However, the land converted to urban development often better responds to the need of profitable revenues than to a genuinely necessary expansion of housing provision, the costs of land acquisition being very low while the sale prices of developed land are very high. As a result, in China the phenomenon of empty buildings or even entire 'ghost towns' is explained by this urgent need to develop new areas without a real demand behind it – though the presence of empty buildings is also due to the use of house purchases as the only option of long-term investment for the Chinese middle class and wealthy families. In terms of energy, the presence of empty constructions is a colossal amount of waste. If we consider construction processes and the life of buildings in figuring the energy wasted in such 'useless' developments, current practices in the urban development in China can be seen as profoundly contradictory with regard to energy security strategies. As we have seen, the life span of buildings is estimated at 20–30 years compared with the far longer life span of European or North American buildings. As for construction processes, issues like life-cycle analysis or the utilisation of materials produced through high energy efficiency processes are still far from being a reality, although some real estate companies are trying to specialise in these aspects (yet there remain few 'flagship' examples).

Chinese cities are therefore currently facing a serious problem of urban sprawl, further backed by an official's performance evaluation system that still focuses on their capacities to boost GDP growth and on their achievements in terms of urban space development, for instance making them privilege the construction of vast modern shopping complexes. Due to this situation, Chinese mayors compete in attracting investments and creating jobs, leading to a further expansion of cities that then creates vicious circles of increasing investments in infrastructures and selling land-use rights to finance their construction. Land speculation thus became a very important phenomenon in China, which also provoked the expansion of projects of very high profitability such as luxurious residential constructions, holiday villas and golf courses. These developments occurred in an unregulated way, where the land of suburban villages is also leased without the backing of land acquisition plans, therefore outside municipal government monitoring (Zhi and Salzberg 2012: 107). Beyond these aspects, another important factor to be considered is the limited voice of urban planners in a rational design of space. Notwithstanding an increasing expertise of Chinese urban planners on the topics of 'sustainable urban planning' – privileging compact cities for instance, with high density, mixed use, etc. – their expertise is often left unheard. Plans are always a matter of negotiation and largely depend on the wishes and 'caprices' of politicians.[24]

The consequences of the expansion of such unregulated cities gave birth to a series of inequalities and dysfunctional aspects that made current urban development patterns unsuitable to considerations of issues of sustainable development, energy conservation or energy efficiency. Beyond urban sprawl, thus the extension of urban peripheries and the development of low-density residential areas, which in turn inherently encourages the use of cars, there is also a series of construction patterns that are clearly unfitting to the purpose of reducing energy consumption and the emission of pollutants. For instance, the construction of wide roads has for the most part been preferred over that of smaller, more suitable ones for walking or cycling. Over the last decades, planning has indeed been pursued to the profit of the development of cars. Another example is the clear division of urban space into functional zoning, distinguishing residential zones from commercial and industrial ones, once more encouraging long commuting trajectories from workplace to residence. More recently, pushed by the profits of the real estate sector and the desires for distinction and safety of the new upper class, residential areas have started to present important disparities. A symbol of this change, the emergence of exclusive luxury-gated communities not only adds up to the already existing social disparities and imbalances in Chinese cities,[25] but also significantly reduces the possibilities to walk around, thus to the possibility of developing less car-oriented patterns of mobility.

China's sustainable urban development of cities and of energy security

China was the first country to ever develop its own Agenda 21 and to reduce it to a local level, subsequently giving birth to national and local programmes to render Chinese cities cleaner and more resource saving. Closer to energy saving and energy efficiency issues, the 'sustainable urban development move' was also added to a new banner, making the development of 'low-carbon cities' (*ditan chengshi*) a national priority. Programmes followed, coordinated by different ministries or the National Development and Reform Commission, where a group of pilot cities was selected on which to draw their sustainable development plans.[26] However, to date, the 'low-carbon cities' discourse still results in a use of empty slogans (Hu 2010), as the discussion provided in this chapter also describes. To summarise, despite existing discourses on the need to make Chinese cities more oriented towards 'sustainable development' (or towards 'low-carbon development', a term that became largely diffused from 2009 onwards), they currently clash with the existing urban development practices and the reality of the environment in Chinese cities.

Beijing's thick and visibly unhealthy air became the emblem of China's environmental crisis, imposing the need to take drastic measures upon the government. In 2013, the latter announced its decision to ban the construction of coal-fired power plants in the areas surrounding the cities of Beijing, Shanghai and Guangzhou (Watt 2013). It also decided to replace four existing coal-fired power plants in Beijing with new gas-fired plants (Standing 2013). The problem hence showed the significant need to rethink the way China currently manages energy supply, while progressively abandoning the use of coal that still remains, which, as we have seen in previous chapters, is the country's main source of energy. Thinking about a strategy of progressive coal replacement, gas started to be considered the primary response. However, as we observed in Chapter 2, the diminishing conventional resources that the country disposes of, together with the current difficulties to tap its unconventional ones that are rich yet difficult to exploit, make the strategy of replacing coal with gas not altogether viable. China's insufficient national gas production already pushes the country to increase its dependence on foreign sources. If the traditional concept of energy security is based on securing oil reserves and designing strategies that are capable of easing the reliance on single producers – especially on instable areas – then as we have seen the issue of dependence is equally valid for gas. Henceforth, in a comprehensive equation regarding energy security, other concepts beyond 'securing supplies' should be taken into consideration.

This chapter points out that if they are not backed by corresponding strategies at the local level, a discrepancy of energy security efforts will occur at the higher level. This disregard of energy management at the city level is not only a Chinese problem, as energy flows have historically been ignored by cities (Raufer 2007: 166). However, in the case of China, until its cities do not enforce energy efficiency policies and promote more energy parsimonious lifestyles, the country's energy quest will not only be a major concern for itself but also for the rest of the world. Moreover, the choice of an energy efficient path should be coupled by an extensive use of cleaner energies if Chinese cities wish to avoid carbon lock-in (at least for those cities that can still adopt significant action) and, above all, to breathe. China has a strong potential for implementation and finance, as is shown by its increasing capacity to adjudge itself foreign assets of energy companies or partnering in exploitation areas of foreign energy sources. If these capacities are directed towards rendering its cities progressively cleaner and more energy performing, a big bulk of the energy security problem could find win-win solutions. Supply would be protected, as would the health of Chinese citizens. In concrete terms, without significant energy prices and fiscal reforms (as well as other measures to enforce the pre-existing instruments at the city level), energy waste and urban sprawl, the two main etiological factors behind energy shortage and irrational energy use, will be likely to persist in the near and far future.

As we have briefly presented in these paragraphs, China's energy security clearly has a local face and is called upon to cope with the pressing needs of combating energy shortage episodes as well as unbearable air pollution. If adequate energy supply (including the limitation of waste) and healthy air are two important elements to be protected by specific energy strategies, an energy security strategy also needs to be equipped by policies at the local level, if not strengthen the already existing ones. In particular, local energy security options encompass the reduction of consumption in buildings and transportation and the enhancement of efficiency in building operations, in the transportation fleet and in the industrial sector, as well as the diversification of energy supply through increased provision of clean energies. From this point of view, the 'local face of energy security' or the 'local dimension of energy security' rhymes with the concept of 'energy transition'.

Notes

1 See for instance Pascual and Elkind (2010) talking about 'climate-smart metropolitan economies', or Brown and Sovacool (2011) referring to the need of grasping the 'benefits of local action'.
2 For an understanding of the overall concept of 'energy transition' see Krause *et al.* (1980).
3 In the administrative division of China, under the provincial level (including the four big municipalities of Beijing, Chongqing, Shanghai and Tianjin) there is the prefectural level, mostly composed by cities and their surrounding rural areas. In China we currently count 284 cities at prefectural level.
4 In 2013, the Chinese urbanisation level reached 53.73 per cent (China National Bureau of Statistics 2014).
5 To give an idea of the huge dimension of this datum, for the World Health Organizations, concentrations of PM 2.5 above 25 micrograms per cubic metre are already considered dangerous for human health.
6 In autumn 2013, when conducting researches in Shanghai city and Jiangsu province, we observed unusual levels of pollution, the air quality index having recorded levels beyond 500, rarely registered in these areas.
7 For instance in 2005, Shanghai introduced a Green Electricity Scheme promoting the use of renewable energy. This initiative – the Jade Electricity Program – using wind and PV energy generated in Shanghai, included mandatory tariffs and subsidies and a voluntary green electricity scheme, covering the incremental costs of renewable energy electricity generation. However, the implementation of the programme has been hindered by different barriers, like inadequate financial incentives, inadequate market and dissemination efforts, insufficient government attention and complicated administrative procedures (Peng 2012: 141–143).
8 Confirmed and highlighted by several interviews with architects and academia questioned on the status of energy efficiency in buildings in China and asking them to compare with the situation in Europe or the US.
9 These buildings do not only consider energy efficiency and energy conservation, but also measures to reduce resource consumption and environmental impacts. Moreover, they are supposed to be safer in terms of the materials used and the indoor atmosphere.
10 Interviews conducted with experts in April 2014 in Shanghai pointed out the fact that the Chinese standard was 'largely inspired' by the American system, the only one at the time to dispose of freely downloadable information on the Internet.
11 It is not possible to provide an exact number of the difference between Chinese energy efficiency standards of buildings with the European average, but this information came from exchanges on energy efficiency in buildings with experts from the university and professionals from the architectural design sector. The experts

all recommended an important revision of the building codes, the 65 per cent target, for instance, still being very far from the European minimum requirements.

12 Different experts in the building sector interviewed during our fieldworks between 2013 and 2014 highlighted this aspect among the weaknesses of the current system.

13 Architects interviewed during our fieldworks underlined how rapid the design process of a building can be, as developers require them to proceed even within short time limits of three to four weeks.

14 In the building design process there are three different design stages. The first stage, called 'schematic design', and the second stage, called 'design development', can be carried by the private architectural cabinet, foreign companies included (up to now still more specialised in energy efficiency measures, the topic of energy efficiency in buildings not commonly taught in Chinese universities). In the third phase, when construction drawings are produced, they can only be done by local design institutes. They are the basis for the construction of the building, and they are produced following the design development plans. However, since Chinese construction codes are not the same as foreign ones and because local Chinese design institutes are under a lot of time pressure, paid according to the number of projects carried out, and also a lack of the necessary knowledge, the original design ideas of (foreign) architects, including energy efficiency measures, are often lost in the process of construction drawings.

15 Real estate advisors interviewed during our fieldwork told us that extra construction costs to reach the levels required by the three-star system can attain 100 RMB/square metre, while the subsidies cannot cover this expense. The subsidy to obtain the two-star levels are set at 45 RMB/square metre, while the three star developers can obtain 80 yuan/square metre. Moreover, the process to obtain the certification can be long and time consuming, finally refraining companies from pursuing the obtention of the label.

16 The most recent datum for China was for 2011. Data can be retrieved in the UN database at: http://data.un.org/Data.aspx?d=WDI&f=Indicator_Code% 3AEG.USE.PCAP.KG.OE (accessed 14 February 2014).

17 In the case of trucks transportation, the recording system attributes energy consumption to the originating sector – the industrial sector as opposed to the 'direct consumer' (the trucks involved in transportation) (IFEU 2008: 6).

18 Data retrieved at the World Bank database of CO_2 emissions for transport at: http://data.worldbank.org/indicator/EN.CO2.TRAN.MT (accessed 14 February 2014).

19 To offer some data on the dramatic pace of diffusion of cars in China, and especially in the two cities, in 2004 both had about two million motor vehicles, but just six years later the discrepancy between car growth trajectories in the two cities became evident, Beijing having 4.8 million and Shanghai only 3.1 million. These data, provided by a research project conducted by MIT Urban Planning, 'Managing cars', show that in 2011 private car owners represented 38 per cent of Beijing households, whereas in Shanghai they were 20 points less. The difference in results is due to specific policy choices taken by the two cities, Shanghai opting much earlier for a car ownership management policy, while Beijing prefers a control over car use and a 'lottery' for car plates attribution, as well as later than Shanghai (after the Olympic games). For info see: http://dusp.mit.edu/idg/project/mana ging-cars (accessed 14 February 2014).

20 This part is mainly based on interviews with experts (mainly urban planners from academia or from the public administration) regarding current patterns in the development of cities in China, carried out by the author to give opinions about the topic of 'sustainable urban development' in China. During the interviews, experts mostly underlined the contradictory practices of the development of cities if taken from a sustainability perspective, current patterns privileging

the development of wide car-oriented roads among other aspects, and pushed by the competition among local officials. We chose the term 'modernisation' because these practices were mainly attributed to the need of urban planners to respond to the expectations of officials to 'modernise' and beautify cities, basically by constructing highways and high-rise buildings and following an urban planning pattern developed in the US. The answer 'We have to follow the American planning style' was often mentioned to explain the reason why cities expand so much and do not privilege spatial configurations that are more bicycle-oriented or pedestrian-oriented.

21 It makes a distinction between urban land, state-owned and use rights being tradable, and rural land, which is collectively owned and under specific restrictions for its conversion (though possible, namely through land acquisition plans backed by city master plans).

22 Many local governments in China established urban development investment companies (UDICs) in order to help get their land transfer revenues and obtain domestic loans to invest in infrastructures.

23 The average sum offered to farmers is 3,000 Yuan per *mu* (local measuring unit corresponding to 0.07 hectares) (Werman 2013), but this is not always guaranteed since land is often seized without compensation (Lin 2009).

24 During our interviews, we were told several anecdotes, sometimes showing irrational requests on behalf of decision-makers, such as constructing central business districts in very small cities where there was no actual need.

25 The issue of social disparities in Chinese cities – especially concerning the matter of residence permits (*hukou*) for rural migrants and the resulting unequal conditions due the difficulties of accessing city services without an urban *hukou* – as well as the environmental issues to which the current development of Chinese cities are confronted, have been purportedly omitted from this paragraph. This choice was justified by the need to look at Chinese urban development problems under an energy security perspective. However, it is important to point out the urgency of these questions that are currently on the list of main pressing problems for the Chinese government. In view of future population movements, the Chinese government acknowledges the need to switch to a more balanced urbanisation process through facing the increasing societal disparities and environmental degradation produced by two decades of blind pursuit of economic growth by local governments.

26 The 'eco-city programme' developed by the Ministry of Environmental Protection or the more recent 'low-carbon cities programme' of the National Development and Reform Commission selected various pilot cities to develop ecological or low-carbon plans and implement related projects in order to attain the targets established by these same plans.

Bibliography

Andrews-Speed, P. (2012) *The Governance of Energy in China. Transition to a Low-Carbon Economy.* Basingstoke: Palgrave Macmillan.

BaeumlerA., Ijjasz-Vasquez, E. and Mehndiratta, S. (2012) *Sustainable Low-Carbon City Development in China.* Washington, DC: The World Bank.

Breheny, M. (1995) 'The Compact City and Transport Energy Consumption.' *Transactions of the Institute of British Geographers* 20(1): 81.

Brown, M. A. and Sovacool, B. K. (2011) *Climate Change and Global Energy Security: Technology and Policy Options.* Cambridge, MA: MIT Press.

Cai, W. G., Wu, Y., Zhong, Y. and Ren, H. (2009) 'China Building Energy Consumption: Situation, Challenges and Corresponding Measures.' *Energy Policy* 37: 2054–2059.

Chinadaily (2014) 'Beijing to Build Seventh Ring Road', *Chinadaily*, 26 June. Available at: www.chinadaily.com.cn/china/2014-06/26/content_17618207.htm (accessed 13 September 2014).

China National Bureau of Statistics (2014) *2013 nian guomin jingji he shehui fazhan tongji gongbao [2013 National Report on Statistics of Economic and Social Development]*. Beijing: China National Bureau of Statistics.

Démurger, S. (2010) 'Editorial – Rural Migrants: On the Fringe of the City, a Bridge to the Countryside'. *China Perspectives* 4. Available at: http://chinaperspectives.revues.org/5331 (accessed 15 September 2014).

Economy, E. (2005) *The River Runs Black: the Environmental Challenge to China's Future*. Ithaca, NY: Cornell University Press.

Fitzgerald, J. (2010) *Emerald Cities: Urban Sustainability and Economic Development*. New York: Oxford University Press.

Fung, E. (2014) 'China Land Prices Pick Up Pace in Fourth Quarter.' *Market Watch* (12 February). Available at: www.marketwatch.com/story/china-land-prices-pick-up -pace-in-fourth-quarter-2014-02-12 (accessed 15 February 2014).

Gheorghe, A. and Muresan, L. (2011) *International and Local Issues, Theoretical Perspectives, and Critical Energy Infrastructures*. Dordrecht: Springer.

Guo, W. and Liu, Y. (2010) 'Zhongguo de wu ai jian pai zhengce: geili ditan chengshi fazhan [Low Carbon City: A Non-Regret Option].' In G. Fan and W. Ma (eds), *Ditan chengshi zai yudong: zhengce yu shijian [Low Carbon City in Action: Policy and Practice]*. Beijing: Zhongguo Jingji Chubanshe, pp. 3–80.

He, J. (2010) 'Zhongguo chengshihua yu tan paifang guanxi shizheng fenxi [China Urbanisation and Carbon Emissions Relations Positive Analysis].' *Xibu luntan [West Forum]* 5: 79–86.

den Hartog, H. (2010) 'Shanghai as Urban Experiment.' In H. den Hartog (ed.), *Shanghai New Towns: Searching for Community and Identity in a Sprawling Metropolis*. Rotterdam: 010 Publishers, pp. 63–80.

Hu, X. (2010) 'Zhongguo chengshi ditan fazhan de wenti yu jianyi [Problems and Suggestions About China's Low Carbon City].' In G. Fan and W. Ma (eds), *Ditan chengshi zai xindong: zhengce yu shijian [Low Carbon City in Action: Policy and Practice]*. Beijing: Zhongguo Jingji Chubanshe, pp. 81–97.

Huang, K. (2013) 'Gov't Grapples With Shortage of Natural Gas', *Caixin*, 29 November. Available at: http://english.caixin.com/2013-11-29/100611459.html?p1 (accessed 10 February 2014).

InstitutfürEnergie und Umweltforschung – IFEU (2008) *Transport in China: Energy Consumption and Emissions of Different Transport Modes*. Heidelberg: IFEU.

Kamal-Chaoui, L. and Cointreau, M. (2014) 'Better Cities, Better Planet: Examples of Governing Against Climate Change From OECD Countries.' In P. Ni and Q. Zheng (eds), *Urban Competitiveness and Innovation*. Cheltenham: Edward Elgar Publishing, pp. 29–38.

Krause, F., Bossel, H. and Müller-Reißmann, K.-F. (1980) *Energie-Wende. Wachstum und Wohlstand ohne Erdöl und Uran. Ein Alternativ-Bericht*. Frankfurt am Main: S. Fischer.

Lam, P.-L. (2005) 'Faiblesses et Perspectives du Secteur Énergétique Chinois.' *Perspectives Chinoises* 88. Available at: http://perspectiveschinoises.revues.org/737 (accessed 15 September 2014).

Larson, C. (2011) 'A Once-Polluted Chinese City is Turning From Gray to Green', *Yale Environment 360°*, 17 October. Available at: http://e360.yale.edu/feature/shenya

ng_a_once-polluted_china_city_is_turning_from_gray_to_green/2454/ (accessed 15 September 2014).

Larson, C. (2012) 'The Cracks in China's Shiny Buildings', *Bloomberg Businessweek*, 27 September. Available at: www.businessweek.com/articles/2012-09-27/the-cra cks-in-chinas-shiny-buildings (accessed 15 November 2014).

Lin, G. C. S. (2009) *Developing China. Land, Politics and Social Conditions.* Abingdon and New York: Routledge.

Ma, L. J. C. (2004) 'Economic Reforms, Urban Spatial Restructuring, and Planning in China.' *Progress in Planning* 61: 237–260.

McKinsey Global Institute (2009) 'Preparing for China's Urban Billion'. Available at: www.mckinsey.com/insights/urbanization/preparing_for_urban_billion_in_china (accessed 15 February 2014).

Naess, P. (2001) 'Urban Planning and Sustainable Development.' *European Planning Studies* 9(4): 503–524.

Niu, W. (2009) 'Zhongguo chengshihua zhanlüe de ditan zhi lu [The Challenge of China's Urbanisation Low-Carbon Way].' In *2050 Zhongguo nengyuan he tan paifang baogao [2050 China Report on Energy and Carbon Emissions]*. Beijing: Kexue Chubanshe, pp. 49–140,

Pan, H. (2011) 'Implementing Sustainable Urban Travel Policies in China.' *OECD International Transport Forum*, Discussion Paper No. 2011–2012.

Pan, H. and Liu, F. (2011) 'Jiaotong fazhan yu ditan chengshi [Transportation Development and Low-Carbon City].' In *2050 Shanghai ditan fazhan luxiantu baogao [2050 Shanghai Low Carbon Development Roadmap]*. Beijing: Kexue chubanshe, pp. 247–280.

Pascual, C. and Elkind, J. (2010) *Energy Security: Economics, Politics, Strategies, and Implications.* Washington, DC: Brookings Institution Press.

Peng, X. (2012) 'Low-Carbon Electricity for Cities.' In A. Baeumler, E. Ijjasz-Vasquez and S. Mehndiratta (eds), *Sustainable Low-Carbon City Development in China.* Washington, DC: The World Bank, pp. 131–145.

Qi, Y. (2013) 'Annual Review of Low-Carbon Development in China (2011–2012).' *Chinese Research Perspective on the Environment.* Special Volume. Leiden and Boston, MA: Brill.

Raufer, R. K. (2007) 'Sustainable Urban Energy Systems in China.' *N.Y.U Environmental Law Journal* 15: 161–204.

Richerzhagen, C., Von Frieling, T., Hansen, N., Minnaert, A., Netzer, N. and Russbild, J. (2008) *Energy Efficiency in Buildings in China: Policies, Barriers and Opportunities.* Bonn: Deutsches Institut für Entwicklungspolitik.

Romano, G. C. (2013) 'On ne Pourra Pas Sortir de l'"Air-pocalypse" Urbaine par des Mesures Administratives.' *Perspectives chinoises* 3. Available at: http://perspecti veschinoises.revues.org/pdf/6671 (accessed 14 February 2014).

Rydin, Y. (2010) *Governing for Sustainable Urban Development.* London: Routledge.

Shui, B. and Li, J. (2012) 'Building energy efficiency policies in China'. *Global Buildings Performance Network/American Council for an Energy-Efficient Economy, Research Report E129.*

Sovacool, B. K. (2011) *The Routledge Handbook of Energy Security.* Abingdon and New York: Routledge.

Standing, J. (2013) 'China's Capital to Replace Some Coal-Fired Heating Plants', *Reuters*, 5 October. Available at: www.reuters.com/article/2013/10/05/us-china-pollu tion-idUSBRE99403O20131005 (accessed 15 February 2014).

Tian, Y. (2013) 'China Renews Electric Vehicle Subsidies Excluding Hybrids', *Bloomberg News*, 17 September. Available at: www.bloomberg.com/news/2013-09-17/china-renews-electric-vehicle-subsidies-without-adding-hybrids.html (accessed 15 February 2014).

US Department of Energy (2006). 'Energy Policy Act 2005 Section 1837: National Security Review of International Energy Requirements' (Report to Congress). Washington, DC.

US Energy Information Administration (2013a) *International Energy Outlook. Buildings Sector Energy Consumption*. Washington, DC: US Energy Information Administration. Available at: www.eia.gov/forecasts/archive/ieo13/buildings.cfm (accessed 15 November 2014).

US Energy Information Administration (2013b) 'International Energy Outlook. Transportation Sector Energy Consumption'. Washington, DC: US Energy Information Administration. Available at: www.eia.gov/forecasts/archive/ieo13/transportation.cfm (accessed 15 November 2014).

V&E Law (2013) 'China Issues New Natural Gas Utilization Policy.' *V&E China Practice Update E-communication* (17 January). Available at: www.velaw.com/resources/ChinaIssuesNewNaturalGasUtilizationPolicy.aspx (accessed 15 February 2014).

Watt, L. (2013) 'New Coal-Fired Plants Ban to Take Place in Beijing, Shanghai and Guangzhou China', *Huffington Post*, 12 September. Available at: www.huffingtonpost.com/2013/09/12/coal-fired-plants-ban-china_n_3914703.html (accessed 15 February 2014).

Werman, M. (2013) 'China Past Due: Land Rights,' 1 May. Available at: www.pri.org/stories/2013-05-01/china-past-due-land-rights (accessed 15 February 2014).

World Bank (2001) *China: Opportunities to Improve Energy Efficiency in Buildings*. Washington, DC: World Bank.

World Bank (2010) *Cities and Climate Change: An Urgent Agenda*. Washington, DC: World Bank.

World Bank (2014) *Urban China – Toward Efficient, Inclusive and Sustainable Urbanization*. Washington, DC: World Bank.

Wu, F. (1999) 'The Rransformation of the Urban Planning System in China From a Centrally-Planned to Transitional Economy.' *Progress in Planning* 51: 165–252.

Yergin, D. (2011) *The Quest: Energy, Security, and the Remaking of the Modern World*. London: Penguin.

Zhi, L. and Salzberg, A. (2012) 'Developing Low-Carbon Cities in China: Local Governance, Municipal Finance, and Land-Use Planning – The Key Underlying Drivers.' In A. Baeumler, E. Ijjasz-Vasquez and S. Mehndiratta (eds), *Sustainable Low-Carbon City Development in China*. Washington, DC: The World Bank, pp. 97–127.

12 Conclusions

Jean-François Di Meglio and Giulia C. Romano

As an intermediatee conclusion, and to contrast the current stage of China's energy security issues with the prevailing situation a decade ago, we would summarise by saying that China has now defined the general 'frame' for the future and that this is much more dynamic than in the past. Instead of considering the global energy solutions as something beyond its reach, China has decided to 'frame' its policy in accordance with a pragmatic spirit and to implement it accordingly. While China's energy mix has only slightly changed, the chapters included in this book clearly indicate the future trends and developments within the context of the 'frame' which is being designed.

In terms of coal, and contrary to the situation that prevailed a decade ago, this resource is now firmly embedded within the energy security equation. This equation used to focus on resources not mastered by China, which basically meant that almost 80 per cent of China's energy security thinking was left 'out of the frame'. China now imports coal, but at the same time it is committed to reducing its reliance on this resource in general and to highly polluting grades of coal specifically. This will take time, however. The new strategic horizon is also oriented towards a diversification of logistics and sources: the proportion of nuclear power in the mix is to increase, as is that of gas, the latter as a follow-up to active pushes vis-à-vis Central Asia and Russia, as well as investments into gas and oil pipelines lowering the reliance on traditional supply routes.

If we revisit the 'scenarios' drawn up in 2007, balanced as they were between a 'business as usual' approach and a full re-orientation of the Chinese energy mix with a greater stress on renewables and cleaner sources of energy, we can see that not a lot has been achieved. This can obviously be explained by China's strong growth in recent years, which has imparted a high degree of resilience to the established scheme, such that any 'improvement' could only be marginal, given the momentum already acquired. Another explanation is to be found in China's domestic policies. Overall, the 'Hu-Wen' period was a very conservative one, and reforms are only becoming manifest with the consolidation of the new 'Xi-Li' duo. Finally, this conservatism, which is more and more openly challenged now, had led to an over-representation of the 'oil lobby', which now gives way to more balanced thinking and forward-looking planning.

Obviously, there is also a far more responsible and integrated posture vis-à-vis energy issues and the challenges posed by the rapid development of the Chinese economy. While still concerned about the potential hindrances that shortages could bring, planning authorities within energy companies (namely the most influential ones, oil companies) or at the government level have been careful in bringing as much diversification in the new sources of energy as possible. Gas for example is an excellent illustration of both the over-whelming strength of oil companies, which are heavily involved in this 'new' resource, and of their desire to counter any risk of supply shortages due to a lack of appropriate sourcing.

However, Chinese energy security policies are evolving in a somewhat similar fashion to certain other policies aimed at managing the ever-growing importance of China in the world but that are not yet fully coping with the emergence of China as a global leader. By the same token, the Chinese monetary authorities have clearly stated that they intend to promote the use of their currency to make it a global player, while they still refrain from accepting the general and international rules governing 'international' currencies, starting with convertibility. This makes China an odd player. In the field of energy, the feeling of being weak and besieged has also given way to a calmer and more balanced attitude vis-à-vis the threat to energy security and the remedy for energy vulnerability.

Energy security, as it has been seen in the various chapters of the book, is still a China-centred affair, as if one country's security could systematically mean insecurity for other countries. China has definitely chosen and managed to enjoy more energy security and has adjusted its organization and govern-ance accordingly. Yet there is still room to integrate this role of 'stage direc-tor' of its energy security within the rest of the world, as could have been demonstrated China showed more enthusiasm, for example, in defending the security of the 'Straits', as opposed to securing its own supply by circumventing the strategic dangers present in the Straits by building its oil and gas pipelines through Burma.

In what follows, we highlight a few of the main findings of our research:

- As a (still) planned economy, China has entrusted the top echelons of the government with the task of watching and regulating the main directions, regulations and allocation of capital in the field of energy planning, supply and energy mix. However, whilst, on the one hand, this centralisation of energy policymaking strives to harmonise energy management, on the other it clashes with the effects of decentralisation of competences to provinces, thereby creating important distortions for the energy markets. As chapter 2 has shown, this is particularly evident in the case of coal, where the central government's attempts to rationa-lise the use and distribution of resources is challenged by the provinces' interests in their own coal industries. But it is also evident when we turn to consider the question of energy transition and urban transition, the planned economy model appearing as an unsuitable approach to trigger the necessary forces to clean up and rationalise the Chinese energy mix.

- Nonetheless, this vertical, top-down approach would not exclude the possibility that competing bodies are overlooked rather than really managed or monitored by 'agencies' (such as NEA and NDRC).
- A short- to medium-term vision, always adaptable, within a framework of established bodies regulating and looking after oil procurement, while nuclear developments have been taken care of by specific bodies, with other sources of energy being under different jurisdictions.
- Energy security is clearly a national security issue and, as such, it is a key component to understanding priorities in Chinese politics. That said, there is room to integrate 'peripheral initiatives' of varying importance and impact. The protracted efforts to improve the environmental impact of energy policies form part of these initiatives, which could be taken at the local, provincial or central level. Many of the chapters in the present work clearly bring out this situation, where the need to make 'clean' energy choices leads to a necessary consideration of the place of each energy resource in the Chinese energy mix, as well as of the patterns of energy consumption.

Compared with what could be said in 2007, when we published our first collective work, the new 'framing' of energy policies, as we define it, has changed priorities. China's weight and voice in the global energy game has strengthened, so its posture is no longer one of a 'challenger' for access to supplies. Yet, and however weighty its influence and important the consideration of its presence in global market (and cartel) processes, China is not part of the 'price-making' process, at least not for oil, the most global of all commodities. Even with respect to international coal prices, China's voice is not absolute. Conversely, China is increasingly becoming a key player in gas prices, at least for the region spanning from Central to East Asia, as well, of course, for commodities which are key to its development (iron ore, copper, etc.).

We therefore need to take a very different approach to the one adopted in 2007 in speculating on the future implementation of energy security policy in China. Any scenario should now definitely factor the international dimension into the equation much more than previously. The growing role of best practice, the concern for air quality, largely influenced by international debates, China's participation in climate debates and finally the country's foray into new spheres of influence, mean that China's approach towards energy security can, in large part, be defined by two key words: pluralism and diversification.

China's policy towards energy security, which had once been obsessed almost entirely with its independence, has obviously kept its perennial stress on securing procurement. This book has also shown that new efforts have been deployed at home in terms of improving energy efficiency, and that China has taken into consideration environment policies and best practices in terms of construction and urban development, although enforcement remains very weak. Equally, the shift away from the age-old energy mix has started. In other words, the new dominant scenario for the future of energy security over the next twenty to twenty-five years should involve the following:

- A confirmation of the ever-growing need for energy. As compared to the current circa 4 billion tce (tons of coal equivalent), which has grown five-fold during the last 30 years, albeit by only 4 per cent over the past year, consumption of energy should be in the range of 5 to 6 billion tce by 2020 and 8 to 10 billion tce by 2030.
- A likely increase in 'cleaner' energy sources to 30–35 per cent of primary energy sources (more nuclear plants, rapid development of nuclear, gas and cleaner coal).

We have to correlate all this to the bilateral agreement reached with the US President at the APEC meeting in Beijing last November 2014, whereby China has clearly stated its commitment to capping its emissions at the maximum level by 2030. Then a clear and coherent picture of things to come can be seen.

Working on Chinese energy policies and describing the correlated landscape might appear to be an endless undertaking. It seemed to us that the current turnaround, confirmed by the official stance taken at the highest state levels by China and the US in late 2014, is a significant basis for such a study. It is our belief that, following the publication of the book, changes may well occur. In this connection, our hope is to have contributed to a proper understanding of the background and roots of what we have chosen to call the 'frame' of China's energy situation and the evolving international context. As a significant player in the global energy game, China's role, vision and influence are complex, yet many elements now exist for a more positive interaction with and understanding by external parties. The conclusions to be drawn show why any positive outlook on China's role should also take into account other complex, and conflicting, elements such as those analysed in the present study.

Index

Note: Page numbers in *italic* refer to illustrations. Page numbers followed by an 'n' refer to notes.

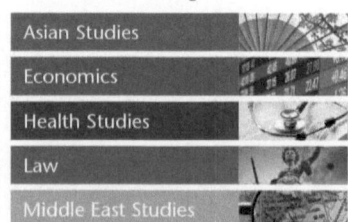